Polearms
OF
Paulus Hector Mair

Polearms
OF
Paulus Hector Mair

David James Knight
and Brian Hunt

Published by Echo Point Books & Media
Brattleboro, Vermont
www.EchoPointBooks.com

All rights reserved.
Neither this work nor any portions thereof may be reproduced, stored in a retrieval system, or transmitted in any capacity without written permission from the publisher.

Copyright © 2008, 2022 by David James Knight and Brian Hunt

Polearms of Paulus Hector Mair
ISBN: 978-1-64837-104-2 (casebound)
       978-1-64837-105-9 (paperback)

Cover design by Kaitlyn Whitaker

TABLE OF CONTENTS

Chapter 1: Introduction 1
   The Life of Paulus Hector Mair 1
   The Manuscripts 4
   Remarks on the Translation 4
   Remarks on the Images 5

Chapter 2: Spear and Shortstaff 7
Chapter 3: Lance and Longstaff 53
Chapter 4: Halberd 79
Chapter 5: Poleax 123
Chapter 6: Various Weapons 157
Appendix A: Latin Glossary 177
Appendix B: German Glossary 183
Appendix C: Literal Translations 187
Appendix D: Register 267

Bibliography 271
About the Authors 273

In hoc libro continentur artis athleticae non solum habitus selectissimi atque approbatissimi, verum etiam vitandi et inferendi ictus subtilior quedam ratio, et scientia, quibus si quis rite usus fuerit, facile in palaestra, equestri concursu, et torneamentis victoriam obtinebit.

Tum etiam complectitur figuras gladiatorum concertantium ex ornatissimas declarationibus habituum adiunctis: addita item sunt torneamenta ante annos quingentos in Germania exercita isti itaque habitus a doctoribus gladiatorum peritissimis excogitati et accersiti, per Paulum Hectorem Mair civem Augustanum non citra magnos labores et sumptus, in honorem principium, heroum, atque artis gladiatoriae amantium nunc demum in lucem edita sunt.

Cod. Icon. 393, 1r

In this book you will find the Art of Athletics, not only its most carefully selected and proven techniques, but also its more subtle strikes, which certain doctrines and teachings have shunned and criticized, yet which, when used properly, will allow you to easily obtain victory in fighting academies, knightly combat, and tournaments.

Also included are elaborate illustrations of athletes fighting according to the techniques therein described . . . .

Thus these techniques, devised and compiled by the most knowledgeable Masters of Athletics, are now brought forth into the world, after considerable labor and expense, by Paulus Hector Mair, burgher of Augsburg, in honor of the Prince, the Hero, and the Lover of the Gladiatorial Arts.

Acknowledgments

David James Knight would like to thank his Latin professor, Dr. Kathryn McKinley, for her superb mentorship; Jackie Ege for her German lessons and help contacting the institutions that hold Mair's manuscripts; the late Mike Cartier (RIP) and his Joachim Meyer study group for their feedback on the translations; the Hunts and Feils for their hospitality; Dieter Bachmann for his assistance with the Hils manuscript; and the good taxpayers of 16th century Augsburg for unwittingly funding the works that made this all possible.

Brian Hunt would like to thank his family for their support of his mad hobbies; Eli Combs for his input on the German language; Stewart Feil and the BYU training group for their help with interpretations and other ideas; and all those who have contributed to growing the knowledge of European martial arts.

Chapter 1

Introduction

THE LIFE OF PAULUS HECTOR MAIR

In 1579, at the age of 62, Paulus Hector Mair was hanged as a thief for embezzling from the treasury of the Imperial Free City of Augsburg, in present-day Bavaria, just north of the Austrian Alps. Thus ended the life of one of the most ostentatious and intriguing figures in the history of German fencing.

Mair was born in 1517, at the dawn of the Protestant Reformation—a tumultuous period that saw the violent decline of the Holy Roman Empire and the forging of Europe's modern nation-states. He was likely born in or near[1] Augsburg, a flourishing commercial center due to its position along trade routes with Italy.[2] His family were burghers (members of the nascent bourgeoisie) who had apparently ridden the tide of Renaissance mercantilism into great affluence.[3] Their wealth must have been considerable, for even as a young secretary in the Augsburg civil service, Mair was known for hosting lavish receptions. This "pathological pageantry"[4] quickly earned him the favor of city officials, who granted him the title of City Bursar (and thus unfettered access to an overflowing treasury[5]) in 1541, at the age of only 24. He took on an additional role as Master of Rations[6] in 1545, and held both positions until his death. As one scholar notes, "[i]t appears that he was held in high esteem, as it would be difficult to explain how his misappropriations, which he could easily commit from his position, and which he employed for further luxuries, went undiscovered for as long as they did."[7]

Although it is tempting to dismiss Mair's success as a product of his privilege, he was also well educated and intelligent. There is no doubt he

---

[1] See Birnbaum 199-203 ("Bucher's results for Frankfurt show that in the 15th century, seventy-seven percent of all Neuburger came from places within ten miles of the City. These findings have been duplicated for a number of other cities").
[2] See Moeller 53 (in cities like Augsburg, "where trade with distant lands flourished … some amassed enormous fortunes").
[3] Birnbaum 272 (noting that, by 1500, many successful burghers had risen into the aristocracy)
[4] Hils 198 (quoting Vgl. Rottinger, Jahrbuch 28, 77-79)
[5] Birnbaum 243 (noting the Augsburg treasury took in over 100,000 guilder per annum from property taxes alone)
[6] Hils 198
[7] Id. Trans. Bachmann

had enjoyed a full humanist education,[8] one that included among other things Classical literature, history, mathematics, Latin, and probably Greek.[9] He may even have studied for a period in Italy, a common practice among those families with the means.[10]

Mair's primary academic interest, however, was the ars athletica (his collective term for the indigenous martial arts of Europe, from wrestling to all manner of fencing). There is no evidence that he himself was an accomplished practitioner, let alone a master, but he was certainly a dedicated enthusiast.[11] He amassed an impressive library on the subject. Works known to have been in his collection include the manuals and sketchbooks of Jorg Wilhalm, Antonius Rast, Gregor Erhart, and Sigmund Ringeck, as well as copies of both the Codex Wallerstein and the Konigsegg-Talhoffer manuscript.[12]

He wrote, perhaps autobiographically, that the study of these arts was "civilizing"[13] and "useful in the rousing of young men."[14] But he also seemed to recognize that they were dying arts. By the 16th century, the face of warfare had shifted from the close combat of the Middle Ages to ranged tactics centered on firearms. The primary weapons of the medieval soldier—sword, spear, and halberd[15]—gradually faded from the battlefield, instead finding sportive roles in fechtschulen (fencing schools roughly analogous to modern martial arts academies).[16] Even on the streets, the sword fell out of fashion,[17] with the rapier becoming the gentleman's self-defense weapon of choice.[18]

Mair must have seen the writing on the wall, for around 1540, he began the staggeringly ambitious task of preserving "the entire art of fencing and its history" in a single illustrated compendium—one that would "exceed [his library] ... in completeness and refinement of lore."[19] He obsessively collected manuals by the old masters and, in true Renaissance fashion, organized their writings into sets of technical instructions. Then he recruited professional fencers to pose for illustrations of each technique. He hired Jorg Breu the Younger—son of the renowned Austrian-trained artist Jorg Breu the Elder—to paint the illustrations in full color, and the finest bookbinder in Augsburg to bind the folia.[20]

His completed work, titled the *Opus Amplissimum de Arte Athletica* (or Ultimate Book on the Art of Athletics), is one of the most impressive of the extant Renaissance fencing manuscripts. Three copies survive in German and Austrian collections.[21] Each consists of two volumes. There are chapters on wrestling, armored combat, knightly tournaments, and fencing with the dagger, dussack, longsword,

---

[8] Moeller 21 ("After 1500 we can observe everywhere that the positions of power ... in large imperial cities were being filled by humanistically trained men.")

[9] Lau et al. 8

[10] Birnbaum 318

[11] Hils 198-199

[12] Hils 200. While Mair did not acquire his own copies of these works before 1544, it is almost certain that he was exposed to them much earlier. When Mair began production on the Opus, he seems to have borrowed heavily from works such as the Codex Wallerstein.

[13] Hils 199 ("sittigend")

[14] Munich, Cod. icon. 393, 2r. ("... atqui affirmant / Athleticam in utilitatem inventutis et omnium virtutum excitationem ...")

[15] Oakeshott 25, 29, 41

[16] Hils 201

[17] Hale 102

[18] These weapons did not disappear overnight, though. In 1570, Di Grassi wrote that the greatsword was a weapon likely to be encountered on the street. Silver, writing in 1598, lists the longsword as a weapon suited to both self-defense and war. In his 1612 manual, Sutor, borrowing from Meyer's teachings of 1570, says the same. As late as the 1620s, the masters in Madrid tested for proficiency with the longsword. And in 1626, Thibault still thought the longsword important enough to show how it was used against the rapier.

[19] Hils 199. Trans. Bachmann

[20] Hils 199. It is interesting to note that, while only two fencers were hired, the artwork depicts a large cast of characters: elderly noblemen, soldiers, landsknechts, Spanish Moors, etc.

[21] A fourth Mair text based upon the notes of fencing master Anthon Rast of Nuremburg and illustrated by Heinrich Vogtherr survives in the Augsburg city archives. Unrelated to the Opus, it is dated to 1553 and contains 110 folia.

rapier, longstaff, quarterstaff, halberd, spear, poleax, a rare form of flail, and unorthodox peasant[22] weapons such as the sickle, scythe, and club (which in some images is just a tree branch). The artwork is exceptionally realistic, a welcome departure from the crude drawings in earlier texts such as the Codex Wallerstein. The text is handwritten, suggesting that the Opus was never mass produced.

The influence of humanist thought on Mair's manual is clear. His instructions are meticulously organized and highly technical. He writes in Latin as well as the vernacular. His Latin, although somewhat irregular, is closer to Classical forms than to the Vulgar Latin of the Middle Ages. His handwriting resembles scripts developed by the Italian humanists of the 16th century. All three editions contain numerous appeals to Classical authorities.

"Most men who are knowledgeable in the history of the Greeks, Romans, and Germans agree on the inventors of these arts, and also the origin, and the time when they were first invented, and in which regions they were used," Mair writes, attempting, like many medieval and Renaissance authors, to trace a direct lineage from his homeland to the ancient world.[23] In his "Prefacio ad Athleticam," or "Preface on Athletics," he credits the development of fencing to the Greeks (specifically the Spartans), whom he claims passed their martial knowledge to the Romans. He writes that the Romans brought fencing to Bavaria during Caesar's Gallic campaigns circa 57 BCE. He then makes the peculiar claim that the halberd (a weapon of 14th century Swiss origin) and the poleax (also a 14th century development) were invented by the Amazons, warrior-women of Greek legend.[24]

Mair may have hoped that inventing ties to the Classical world would "civilize" fencing in the eyes of his peers.[25] Or he may have actually believed these founding myths. His source, he claims, was a "man most learned of truth" named Johannes Aventinus[26]—a fiercely nationalistic Bavarian historian who lived from 1477 to 1534 and who, as one scholar puts it, "did not flinch from an occasional departure into pure fact."[27] Mair also mentions "many others [besides Aventinus who] have written ... concerning these same things,"[28] but he does not name them.[29]

In any event, the production of the Opus was an enormous financial burden for Mair. Between the Opus, his library, his "great household," and his extravagant lifestyle, Mair eventually strained his own finances to the point that he began pilfering from the Augsburg treasury.[30] It is unknown when this began or how long it continued, but after a dispute with an assistant "of unfriendly disposition" circa 1579, an audit of Mair's books revealed his embezzlements. He was arrested, tried, and sentenced to death by hanging, aged 62 years. His private holdings were probably seized by city officials in an attempt to recover their losses. His library was scattered at a public auction.[31]

---

[22] See Munich, Cod. icon. 393, 2r (referring to peasantry as "Qui humili de stirpe nati sunt," or "Those born of low stock")

[23] Munich, Cod. icon. 393, 2r. ("Porro historici probati ex Gracis, Latinis, atque Theutonicus plerique / consentiunt de inventoribus istius artis, nec non origine, et quo tempore ea / primum sit inventa, quibusque regionibus fuerit in usu.")

[24] Munich, Cod. Icon. 393, 14r, 14v

[25] By the Renaissance, historical invention had become firmly entrenched in the European literary tradition. Earlier writers—such as Chaucer in his Troilus and Criseyde, written circa 1385 CE, and Geoffrey of Monmouth in his 12-century Historia Regum Britanniae—exercised considerable creative license in connecting their texts to Classical civilizations.

[26] Munich, Cod. Icon. 393, 15r ("... Aventinus veritatis studiossimus ...")

[27] Borchardt 166

[28] Cod. Icon. 393, 14v ("... non citra magnam rei meae familiaris iacturam, omnes cognoscent ...")

[29] Aventinus, in turn, may have borrowed from the wildly fanciful scholarship of Sigismund Meisterlin, whose 1456 *Chronographia Augustensium* places the founding of Augsburg during an Amazon conquest of Bavaria at least 500 years before the founding of Rome. Meisterlin's theories were later echoed, with some modification, circa 1469 by an anonymous Augsburg chronicler, among others, but in reality, the city was founded in the 3rd century by Caesar Augustus.

[30] Hils 198

[31] Hils 198

THE MANUSCRIPTS

MS C93 & C94

All of the images in this translation, as well as most of the German transcriptions, are taken or adapted from the Dresden edition of Mair's Opus. The first volume, MS C93, has 244 folia; the second volume, MS C94, has 328 folia. It is written entirely in Early New High German using a florid blackletter script.[32] The original paintings are in full color; the inking is black. The Sachsische Landesbibliothek, or Saxon State Library, in Dresden, Germany, currently holds this work.

Codex Vinobonensis 10825 & 10826

All of the Latin—and a handful of the German—transcriptions in this translation are taken from the Vienna edition of Mair's Opus. The first volume, Cod. Vinob. 10825, has 270 folia; the second volume, Cod. Vinob. 10826, has 343 folia. It is written in both Latin and Early New High German; the Latin script is in a clear but heavily abbreviated humanistic form, while the German is in a nearly illegible blackletter. The original paintings are full color; the inking is black. The Osterreichische Nationalbibliothek, or Austrian National Library, in Vienna currently holds this work

Codex Icon. 393 & 394

Several short passages were transcribed and translated from the Munich edition. The first volume has 309 folia; the second has 303 folia. It is written entirely in Latin using a crisp humanistic script. This edition was sold to Duke Albert V of Bavaria for 800 florins in 1567. The Bayerische Staatsbibliothek, or Bavarian State Library, in Munich, Germany, currently holds this work.

REMARKS ON THE TRANSLATION

Although we have included a translation of the German text in Mair's Opus, our work is based primarily on the Latin. Creating an accurate and idiomatic translation from what was itself a translation of Mair's vernacular was no small task. The Classical Latin taught in universities today is the heavily inflected language of the Romans, with a complex grammatical structure and seemingly infinite permutations for every noun, adjective, and verb. It is no exaggeration to say that a single verb may be conjugated in well over a hundred different ways, and a noun or adjective declined in at least a dozen. Even in its purest form, Latin's sheer complexity causes many scholars to stumble.

Thus it is no surprise that Mair's Latin, a thousand years removed from the last days of the Roman Empire and influenced by the corrupt Vulgar Latin of the Middle Ages, presents a number of challenges to the modern translator. Initially, we worked only with the Vienna manuscript, which required us to wade through Mair's heavily abbreviated, sometimes illegible script. For the first few weeks, we each prepared separate transcriptions and compared the two to resolve ambiguities. As we became more fluent in Mair's handwriting, I prepared the Latin text myself. Meanwhile, Brian deciphered the German in the Dresden edition, occasionally cross-referencing with the Vienna's less legible German. Once we acquired a copy of the Munich edition (which is written in a very clear hand), I was able to check the Latin transcriptions for any undiscovered errors.

We then repeated this process with our translations. Until we had a firm grasp of Mair's Latin, this was very challenging. It was often difficult to identify exactly which word he was using and how he intended to use it. Like many Renaissance authors, Mair regularly employs unusual noun forms, such as *adversariis* for the dative or ablative plural of *adversarius*, "adversary." He also occasionally confuses verbs such as *adversare*, "to direct," with deponent verbs like *adversor*, "to oppose." It was often only by comparing the independent Latin and German translations that we were able to sort out the idiosyncrasies of Mair's writing.

After we translated our respective sections of the Opus, I then combined the two into a modern "plain English" translation. Because Mair's grammar does not always conform to Classical conventions, I took some liberties where a strictly literal translation would have rendered the text

---

[32] The script appears to be a form of Schwabacher, as it is more rounded than Fraktur. Schwabacher was heavily used during the Protestant Reformation.

unnecessarily wordy or ambiguous. For instance, Mair tends to introduce a condition and its result with a future indicative and present active subjunctive verb, respectively. This in effect combines the rules of Simple Fact Future and Future Less Vivid conditional phrases. I have, in the interests of fluidity, translated all such constructions as Simple Fact Future phrases, as in "if he does this, then you will do that."

Even when Mair's grammar is correct, it is sometimes awkward. The Latin perfect passive participle, for instance, is most literally translated into English as "having been [action]ed," so a participle such as *porrectus* technically reads "having been extended." However, I have rendered such forms as simple past participles when they serve a more adjectival than verbal function; returning to the previous example, *porrectus* becomes "extended." I also translate most ablative absolutes as gerunds, "by doing [the action]", or Ablatives of Time, "once [the action] has been done; while doing [the action]." In Renaissance Latin, intensive pronouns such as *ipse* were commonly used as demonstratives, so I often remove the emphasis and supply possessives for clarity; hence, "strike at the enemy's chest from the right side to his own right side" becomes "strike at the enemy's chest from your right side to his right side." Even Mair's adverbs, though immune from inflection, are cumbersome at times. *Vero,* for example, translates to "yes, in truth, certainly, to be sure, however." Oftentimes, I omit it to prevent redundancy, and when I feel that a particular instruction should be emphasized, I render the parent verb as an infinite, "be sure to do [action]," instead of the cumbersome "to be sure, you will do [action]."

In all translations, modern punctuation was supplied. Authors' notes appear in parentheses in the final translation, brackets in the appendices. Macrons (horizontal accent marks placed over long vowels) were not used in Mair's time and thus do not appear in our transcriptions.

There are also several fencing terms of art that simply did not translate well from 16th-century German to Latin, much less to English. Because these terms—oberhau, abnemen, and the like— are familiar to modern students of historical European martial arts, we have decided to keep them in the original German for the final translations. Their literal meanings may be found in the appendices.

REMARKS ON THE IMAGES

The illustrations in this text are taken from the Dresden codices, C93 and C94, which we felt were superior to the Vienna edition in terms of both artistic realism and digital image quality. (While the artwork in the Munich edition of Mair's Opus is arguably the best of the three, sadly, it is not available for reproduction.)

Since their creation more than 500 years ago, the illustrations in the Dresden volumes have been copied into a number of different media, losing their color and much of their original detail in the process. The original folia were produced in the workshop of Jorg Breu the Younger, a Bavarian artist of some renown, and presumably remained in Mair's library until his execution. After passing through an unknown number of private collections, they were eventually acquired by Saxon State University, photographed, and stored in microfiche form. The black and white microfiche images were, in turn, converted to high-resolution, high-contrast PDF files before eventually making their way onto the Internet as JPG files. Due to legal and cost constraints, we were forced to work exclusively with those JPGs (which, when this book was first published, were very low resolution; much better images are now available online).

In some of the images, the position of the hands, feet, or weapons were impossible to distinguish from shadows; in others, the ink had simply flaked off or faded away over time. To assist the reader in interpreting Mair's techniques, I have tried to digitally restore all the important details. I began this tedious process by removing all visible dust, scratches, water stains, and other "noise" from the PDF files using photo editing software. I also removed the original ground shadows—which were so dark in many images that they obscured the fighters' foot positions—and replaced them with horizontal hatches. Then, after studying corresponding images in the Vienna and Munich editions, I redrew faded or flaked lines, straightened or rescaled portions of each figure that had been distorted by the curvature of the original pages, and added shadows or lightened areas where necessary to clarify body or weapon positions.

There are also several images in this text that do not appear in the Dresden codex. They correspond to the handful of techniques translated solely from the Vienna manuscript. Although at first glance they may appear to be originals, they are in fact composites I pieced together from images in C93 or C94. To create them, I first studied the images in the Vienna edition (which we were unfortunately unable to reproduce due to prohibitive licensing fees), and then began an exhausting search through the Dresden manuscript for similarly positioned figures—often finding only an arm here, a leg there, or perhaps a weapon that could be flipped or rotated into the correct posture. At times, I was forced to simply draw some parts from scratch. In all instances, I took great care to mirror the style of the original artist. Composite illustrations are marked with an asterisk in the Register found in Appendix D.

The result, we believe, is a text with images that retain the feel of the original artwork while more clearly corresponding to Mair's instructions, thereby allowing readers to interpret and master each technique on their own.

—David Knight, October 2006
(revised December 2021)

Chapter 2

Spear and Shortstaff

"**HASTULA** deinde sequitur originem suam ab Romanis et Persis ducens, / 18 habet habitus praestantissimos" (Cod. Icon. 393, 14r, 11–12).

"The **SPEAR** follows next, drawing its origin from the Romans and Persians. [This chapter] has the 18 most excellent techniques." Four additional techniques exclusive to the Vienna manuscript are also included.

The weapon in this chapter appears to be a staff roughly 5 to 6 feet in length. However, Mair identifies it by the Latin word *hastula*, a diminutive form of *hasta*, "spear." Mair's German word for this weapon, *spiess*, also means "spear."

The spear, a wooden shaft roughly the length of a man, with a sharpened point or a blade mounted at one end, is among the earliest weapons devised by man. It is primarily used for thrusting, though a spearhead is capable of slashing and either end may be used for striking.

Mair's word choice, his reference to the Romans and Persians, both cultures known to have fought with short infantry spears, and his distinctly linear footwork befitting a weapon used in phalanx warfare all suggest that this chapter deals with spear techniques. We conclude that the shortstaff may be used as a training weapon for the spear, and that the two are largely interchangeable.

There is a deliberate order to Mair's manual. The shortstaff chapter sets the foundation for all of his polearms, and thus most of the techniques in this chapter can be applied to the lance, longstaff, halberd, and poleax. There are also similarities between shortstaff fencing and halfswording with the longsword, and thus it may be advantageous for the reader to reconcile the techniques in this chapter with his longsword studies.

Mair's terms for the parts of the shortstaff are as follows: the "forward" or "long point," often known by other authors as the "head" in the context of spears; the "back" or "short point," elsewhere known as the "butt;" and the "shaft." Though Mair does not specify the type of wood, ash was typically used for the shaft.

SHORTSTAFF 1

Cod. Vinob. 10825, 154r

Primi duo superiores ex primo congressu contactus de latere dextro.

Ad huc congressum hasta gubernanda athletice, hoc modo te compones, contra adversarium pede dextro procedes, atque superne in primo / congressu de humero tuo dextro contingas, manu dextra exterius eius hastae, eoque contactu observabis diligenter, num firmiter vel contra, hastam / manibus contineat. / Si fortiter restiterit adversarius, sinistro insequitus pede, et hasta ab ipsius hasta circumvoluta ex latere eius / sinistro dextrum versus, nec non versus visum adversaris pungito.

Sin autem eadem ratione contra te is fuerit usus, teque, in contactu primis congressus / consistente adversus adversarium, dextro pede praefixo, manumque dextra in media hasta continueris priorem, mucrone anterioris / eius conatum eludes in dextrum latus. Interea pede sinistro introgreditor, et mucronem posteriorem contra visum eius impellito. Verum / si conatum istum averterit, dextro consequitus, caput adversaris mucrone hastae tuae superioris ex hastarum tactu affliges.

Ceterum si / caput tuum superne is appetiverit, pedem sinistrum refertum, atque mucrone anteriori excipito, statim vero post pectus eius impulsu adgreditor / Hoste id avertente, impulsum rursus arripe, et dextrum euis brachium adpetito. At interim ab eo retrorsum in tuis tamen / defensione tute recedes.

C93, 183r

Die ersten zwaii obern anbinden von der rechten seiten.

Item schicth dich alsso Mit disem zufecten Inn der stanngen / trit mit deinnem rechten fuoß zu im hinein und bind Im oben von deiner rechten Achsel mit der rechten / hand außwenndegan sein stanngen Inn dem anpünd empfind ob er waich oder hert In dem anpund seu / Ist er hert gegen dir und helt dir starctn wider so volg mit deinnem lincken schennctel hinnach Indes wechsel / im durch von seinner lincken auf sein rechten seiten und stich im zu seinnem gesicht.

Sticht er dir allso nach / deinnem gesicht unnd du auch gegen Im In dem anpünd steest deinnen rechten fus furgeseßt dein rechte hand / furgewenndt mitten Inn deinner stanngen so seß Im das ab mit deinnem vordern orth auf dein rechten seitten / trit mit deinnem lincken schennckel hinnein und wind Im deinem hindern Orth zu seinem gesict sest / er dir das ab so volg mit deinnem rechten schennckel hinnach und schlag Im mit deinem obern orth auß dem / anpünd zu seinnem haüpt.

Schlecht Er dir allso oben nach deinnem haupt so seß deinen lincken schenckel / zu ructn unnd verseß Im das mit deinnem vordern orth Indes stoß Im zü seiner prüst verseßt er dir das so / zücth deinnem stoß wider unnd stoß Im nach seinnem rechten Arm wennd dich dannt zü rüctn von Im / Inn güte versaßung.

The first two upper binds from the right side.

For this bind, position yourself by athletically wielding the staff as follows: advance toward the enemy with your right foot, bind him above your right shoulder with your right hand on the outside of his staff (as depicted in the illustration), and during this bind, feel whether he is holding his staff strongly or weakly. If he is strong, then advance with the left foot, change him through from his left to his right side, and thrust into his eyes.

If he employs this technique against you during the first bind while you are standing with the right foot leading and your right hand forward, holding the middle of the staff, then parry his attack downward with your front point on the right side, step to the inside with your left foot, and thrust your back point into his eyes. If he sets that attack aside, then pass forward with your right foot and strike his head from the bind with your front point.

If he attacks your head from above, then bring your left foot back, ward him off with your front point, and be sure to immediately thrust into his chest. If he parries, then thrust again, aiming for his right arm. At this point, you may safely withdraw from him in a strong guard.

Shortstaff 2

Cod. Vinob. 10825, 154v

Primae duae inferiores ex congressu primo hastarum collisiones de latere sinistro.

Si praedicto habitu commode uti volueris, sinistrum pedem praefiges, supra sinistrum humerum hastam tenebis, ita tum ut manus sinistra / sit praeposita. Si itaque contra adversarium hoc modo constiteris, isque vicissim contra te in primo hastarum contactu inferioris, tum / dextro pede intro concedes, atque de intrinsecus supra brachium sinistrum eius extrinsecus pungito, verum in ipsa punctione, mucronem / in pectus adversariis impellito. Et si eum eluserit, rursus pede sinistro consequitor, et inflecte posteriorem mucronem hastae tuae contra / hostis visum.

Sin vero idem contra te usurpet adversarius, te sinistrum praeponente pedem, punctionem eius avertito, dextro / insequitus, mucronem anteriorem in visum hostis impellito. Verum si exciperit, hastam circumflectendo, versus latus eius sinistrum / eam protrudito. Sin rursus conatum tuum eluserit, sinistro pede insequitor, atque confestim anteriori et posteriori mucrone visum adversariis / adgreditor.

Verum si is habitu eodem usus fuerit contra te, tum avertes hastarum contactu ex tuo latere utroque. Interim / vero insequutus pede dextro, in visum eius mucronem hastae propellito. Caeterum hostis id si exciperit, pede dextro reducto, hastam / manu conversa capiti infliges hostis, subito autem rursus hastam manu sinistra adpraehende, atque faciem tuam accurate observato defensione / eius tutissima.

C93, 183v

Die erste zway undere anbinden von der lincten seiten.

Item schicth dich allso mit dem zufechten / stee mit deinnem lincken fuoß vor und halt dem stanngen auf deiner lincken achsel dein lincke hannd / furgewendt stastu dann also gögen Im under auch gögen dir Inn dem unndern anpünd so trit mit deinem / rechten schennctel hinnein und stich Im von In wendig auß wendig ober sein lincken arm Inn / dem stich Im den Orth zü seinner prüst verseßter dir das so volg mit deinnem lincken schenctel wider / hinnach unnd wind Im deinnen hindern Orth zü seinem gesicht.

Sticht er dir also nach deinem gesicht / und dü mit deinnem lincten fuoß vor steest so seß Im dem stich ab und trit mit deinnem rechten fuoß hinnein / Indes stoss im deinnen vordern orth zu seinnem gesicht seßt er dir den ab so wechsel Im indes durch / und stoß im zu seiner lincken seiten seßt er dir das ab so volg mit deinem lincken fuoß hinnach und wind / Im Inn des behennd zwifach mit deinnem vordern und hindern ort zü seinem gesicht.

Windt er dir also zwifach / ein so seß im das ab mit deinnem anpünd von deinen baiden seiten Indes volg mit deinnem rechten / schennctel hinnach unnd stoß in zü seinem gesicht verseßt er dir das so zeüch deinnen rechten schencte / zü rüctn unnd schlag Im mit deinner stanngen mit verterter hand nach seinnem haüpt Indes greiff / mit deinner lincken hannd widerumb Inn dein stanngen und hab deiner gesichts acht mit güter versaßung.

The first two lower binds from the left side.

If you want to use the aforementioned technique, then position yourself in this way: set the left foot forward and hold the staff above your left arm, with the left hand leading. If you thus stand against the adversary—and he against you—in a low bind (as depicted in the illustration), then pass to the inside with the right foot and thrust from the inside to the outside, over his left arm, making sure to drive your point into his chest. If he parries this, then advance again with your left foot and wind the back point of your staff into his face.

If he thrusts toward your face in the same way, and you are leading with the left foot, then set his thrust aside, step with the right foot, and thrust your front point into his eyes. If he parries, then by winding the staff around, thrust into his left side. If he sets your attack aside, then advance with the left foot and immediately wind into his eyes with the front and back points.

If he does the same thing to you, then parry him from the bind on both sides, pass forward with your right foot, and thrust into his eyes. If he parries this thrust, then bring your right foot back, invert your right hand, strike his head, quickly seize his staff with your left hand, and be careful to protect your face with a strong guard.

Shortstaff 3

Vienna 155r

Secundi hastarum contactus superni duo de latere sinistro.

In contactum supernum de latere sinistro formatum, hasta rite gubernando ita te accommodabis. Si adversarius contra te in praedicto / habitu, scilicet contactu hastarum de latere sinistro constiterit, dextro introgreditor pede, atque diligenter observato, num hastula / firmiter vel minus contineat in primis congressus tactu. Si infirmus reperietur, dextro statim consequeris, et pungendo faciem / eius adpetito. Sin vero conatum eum effecerit irritum, confestim circumflectenda hasta ex latere tuo dextro in latus adversariis / dextrum eam adaptabis.

Verum si is idem contra te usurparit, et tu utrissim in hastarum contactas habitu consistas sinistro / praefixo, conatum eius excipies, verum dextro confestim insequitor, atque visum hostis supra dextrum brachium eiusdem hasta tua / adpetito. Excipiente id adversario, mucrone in eius latus dextrum circumflectendo accommodato, eum latus dextrum impellito. Hostis si id adverterit, pede sinistro consequitor, et intra brachium adversariis utruque versus faciem hastae tuae mucronem posteriorem / inflectendo propelles.

Ceterum ea ratione impulsus adversarius si utatur, dextrum referes pedem retrorsum, atque impetum eius / mucrone hastae tuae anterioris avertere memineris, interimque mucronem posteriorem vicissim in visum eius impelles, dextro rursum / consequitus, hastam manu utraque capiti eius infligito, eoque habitu ab adversario discedes, in tuta tamen tuis defensione.

Dresden 184r

Die anderen zway obern anpinden von der lincten seiten.

Item schicth dich allso mit dem zufecten in den obern anpund / von deiner lincken seiten stat er dann auch allso gegen dir Inn dem anpund von seiner lincken / seiten so trit mit deinnem lincten schennctel hinnein Indes empfind ob er hert oder waich sen Inn dem / anpund Ist er waich so trit mit deinnem rechten schennctel hinnach und stoß Im nach seinem angesicht. Nimpt er dir das ab so winnd dich Indes durch von deinner rechten aüf sein rechten seiten.

Windt / er dir allso nach deinner rechten seitten und dü auch gegen Im steest In dem anpünd deinen lincken fuoß / furgeseßt so verseß Im das in dem trit mit deinem rechten schennctel hinnach und stoß im nach seinem / gesicht uber seinem rechten arm hinein verseßt er dir das so wind dich durch mit deinem ort aüf / sein rechten seiten und stoß Im mit deinnem Ort nach seinner rechten seiten seßt er dir das ab so volg / mit deinnem lincken schennctel hinnach und wind im deinnem hindern ort zwischen seinnen baiden / Armen hindurch zü seinnem gesicht.

Windt er dir allso zü deinnem gesicht so seß deinnen rechten schenctel / zü rückn unnd seß Im das ab mit denem vordern orth Indes wind Im deinnen bindern ort auch zü / seinnem gesicht volg mit deinnem rechten schennctel wider hinnach und schlag Im mit der halben / stanngen mit baiden hennden nach seinnem haüpt zeüch damit ab Im gutter versaßung.

The second two upper binds on the left side.

Position yourself for an upper bind on the left side by wielding the staff thusly: if the adversary stands against you in this bind, then step to the inside with the left foot[1] (as depicted in the illustration) and feel whether he is stronger or weaker than you in the bind. If he is weak, then immediately pass forward with your right foot and thrust into his face. If he parries, then instantly adapt by winding the staff from your right side into his right flank.

On the other hand, if he winds toward your right side in the same way, and you are standing in the bind with your left foot forward, then parry his attack, immediately advance with the right foot, and thrust your staff over his right arm into his eyes. If he parries your thrust, then wind the point around to his right side and thrust into his right flank. If he sets you aside, then pass forward with the left foot, wind your back point between both of his arms, and thrust into his face.

However, if he thrusts toward your face in this way, then bring your right foot back, set him aside with your forward point, and wind the back point around to thrust toward his eyes; then, follow again with the right foot, strike his head with both hands by half-staffing, and withdraw from him in a strong guard.

---

1. The Latin erroneously reads *dextro . . . pede*, "with the right foot."

Shortstaff 4

Cod. Vinob. 10825, 155v

Duae collisiones hastarum inferiores de latere dextro.

Quum infera hastarum collisione recte uti volueris, dextrum pedem praepones, manumque dextra recta, iuxta mucronem anteriorem, et inde sursum / hastam convertes, ut manus dextra iuxta femur adpareat, atque faciem adversariis adpetes.

Caeterum si is idem contra te conabitur, te / quoque in habitu tactus hastarum consistente, dextrumque praeposueris, mucrone anterioris ipsius punctionem avertere non dubites, teque ita / componas, quasi adpetere hastae impulsu eius crus dextrum velis. Verum sinistro consequutus pede, mucronem posteriorem versus visum vel / pectus eius coniicies.

Porro idem hoste usurpante, dextro recedes retrorsum, atque eum conatum adversariis, loco hastae posterioris eludes. / Sed rursus pede dextro introcedere non oblivisceris, et mucronem anteriorem in visum eius propellito. Sin vero te abegerit hostis, hastam / ab eius hastam circumflectes, atque consequutus sinistro, praedicto mucrone versus latus eius dextrum pungito.

Et si te adversarius ea ratione / fuerit adgressus, id avertito, confestim vero pedem dextro consequutus, rotando hastam, mucronem utrumque dupliciter vultus eius / impingito. Verum si eos eliserit, dextrum reducere memineris pedem, atque capiti eius hastam infligito. Curam vero eam habebis / in discedendo ab hoste, ne reperiaris nudus.

C93, 184v

Die anderen zway undere anpinden von der rechten seiten.

Item schuth dich allso Mit dem zufechten Inn den / unndern anpund seß deinnen rechten fuß für dein recht hand gestrackt vornen bey dem ort Indes / wind dich auf dein rechte hand auf die hufft unnd stoß Im zü seinem gesicht.

Stoßt er dir allso / nach deinnem angesicht und dü auch gögen im in dem anpund stast deinnem rechten fuoß furgeseßt / so seß im das ab mit deinnem vordern Orth und thü sain wollest dü im zü seinem rechten / schennckel stoßen Indes trit mit deinnem lincken fuoß hinnach und wind im deinen hindern / orth zü seinnem gesicht oder der prüst.

Windt er dir also zü deinem gesicht so trit mit deinnem rechten / fuoß zü rückn und seß im das ab mit deinem hindern orth Indes trit mit deinem rechten fuoß / wider hinnein und stoß im mit deinnem vordern orth zü seinem gesicht versßt er dir das so wechsel / Im durch mit deinnem orth unnd volg mit denem lincken schenckel hinnach stoß im damit auf / sein rechte seitten.

Stost er dir also zü so seß im das ab inn dem volg mit deinem rechten schennckel hinnach / und wind im mit diennen baiden ortern zwifach zu seinem gesicht seßt er dir das ab so zeöch / deinen rechten fuoß zü rückn und schlag im nach seinem haupt wend dich damit zü rückn in gute / versaßunng.

The second two lower binds on the right side.

To correctly execute the lower bind, position yourself thusly: place your right foot forward, with your right hand near the front point (as depicted in the illustration), then immediately bring your right hand to your hip, rotate the staff upward, and thrust into the adversary's face.

However, if he attempts this same thrust to your face while you are standing in the bind with your right foot forward, then set his thrust aside with your front point and position yourself as if you are going to thrust into his right leg. Instead, immediately pass forward with the left foot and wind your back point into his eyes or chest.

If he winds toward your face in the same way, then step backward with your right foot, set his thrust aside with your back point, pass to the inside again with your right foot, and thrust your front point into his eyes. If he parries, then change him through, follow with the left foot, and thrust your front point into his right flank.

If he thrusts toward you in this manner, then set him aside, immediately pass forward with your right foot, and strike his face twice—once with each point—by winding your staff around. If he parries both blows, then bring your right foot back and strike him in the head. Be careful not to leave yourself open as you withdraw.

Shortstaff 5

Cod. Vinob. 10825, 156r

Habitus aversionis contra duplicem ictum medium.

Ad habitum praenominatum hasta gubernanda more athletico, hoc modo te compones, pede sinistro intro procedes, hasta tua humum / parte priori respiciat, manus vero dextra supra caput recta sit constituta, interimque si sursum hastam rotando converteris, et / manum dextram femori iniunxeris, mucronem anteriorem vultiis hostis infigito.

Verum si adversarius idem fuerit conatus, tuque in / habitu duplicis ictus mediis contra ipsum constiteris praefixo pede dextro, tum eius conatum parte hastae anterioris avertito, confestim / vero pede sinistro insequutus, mucronem hastae posteriorem in faciem eius impellito. Sin autem ab eo habitu se tutum reddiderit, / tum inferne circumflectendo, eius brachium dextrum quassabis. At si rursus se liberaverit, dextro insequere intro, atque mucrone / longo eius visum sauciato. Quum vero eum impulsum hostis animadverterit, eluseritque, dextrum pedem referto, et impulsu pedem adversaris / sinistrum adgreditor.

Sin autem is te eadem ratione laedere impulsu conetur, non avertes eius impetum, verum celeriter pungendo / eius visum sauciato, inde vero ab adversario circumspectius recedes.

C93, 185r

Ein absetzen gögen ainnem zwifachen mittelhaw.

Item schicth dich allso mit dem zufechten indes absetzen / trit mit deinnem lincken schenckel hinnein dein stanngen aüf der erden dein rechte hand gestrach ob deinnem / haüpt indes wind dich aüf dein recte hannd auf die hufft und stoß im zü seinem gesict.

Stost er dir allso / nach deinnem angesicht und dü gögen im stest inn dem zwifachen mitelhaw deinnen rechten fuoß furgefeßt / so seß im seinnen stoß ab mit deinnem vordern orth Indes trit mit deinnen lincken schennckel hinnach und / stos im deinnen hindern ort zü seinnem gesicht verseßt er dir das so wechsel im unnden durch und schlag im / nach seinnem rechten arm verseßt Er dir das furbass so trit mit deinnem rechten schennckel hinnein unnd / stoß im mit deinnem stanngen orth zü seinem gesicht wirt er des stoß gewar und verseßt der den so zeüch deinen / rechten schenckel zü rückn und stoß im nach seinem lincken schennckel.

Stost er dir also nach deinnem linctken / schennckel so verseß im das mit sonder stost im behennd nach seinnem angsicht indes wend dich damit von / im zü rückn inn gute versaßung.

A parry against a double *mittelhaw*.[1]

To execute the aforementioned parry, position yourself as follows: step to the inside with your left foot and place your front point on the ground, with your right arm extended above your head (as depicted by the left figure), then wind the staff upward, bring the right hand to your hip, and thrust your front point into the enemy's face.

If he attempts this same thrust to your face while you are standing with your right foot forward during a double *mittlehaw*[2] (as depicted by the right figure), then set his thrust aside with your forward point, simultaneously pass forward with the left foot, and thrust your rear point into his face. If he parries your attack, then change your staff underneath his and strike his right arm. If he parries again, then step to the inside with your right foot and thrust the long point into his eyes. If he anticipates your thrust and parries, then bring your right foot back and thrust into his left leg.

If the enemy thrusts toward your left leg in the same manner, then do not parry his attack; instead, quickly thrust into his eyes, and then carefully withdraw from him in a strong guard.

---

1. German, "middle-strike"
2. Note that there is no description of the "double middle-strike," nor is there an explanation for the right figure's crossed arms and seemingly impossible hand position.

SHORTSTAFF 6

Cod. Vinob. 10825, 156v

Dextrum et sinistrum propugnaculum seu castrum.

In castrum ex latere dextro ita te adaptabis, praeponito dextrum pedem, de latere dextro hastam contineas in loco eius firmiore. Et si / adversarius contra te de latere suo sinistro vicissim in loco fortiori consistat praefigens sinistrum, et utrique invicem hastas ex primo / congressu contigeritis, te componito, tamenque visum hostis velis adgredi impulsu. Sed ex ea fraude, circumflectendo hastam rotabis ex / latere eius dextro versus sinistrum, atque insequutus pede sinistro, pectus adversariis punctione adgreditor. Sed si hostis averterit / impulsum, dextro subito insequere, et mucronem hastae posteriorem intra adversariis brachium utrumque in visum impellito.

Sin vero / idem is usurparit, dextro pede intro procedes, atque mucrone posterioris impulsum eius avertes, interim vero longa mucrone / rotando hastam ab eius hasta transmuta, atque vicissim eam in eius faciem propelles. Verum si id averterit adversarius, tum si / sinistrum praefixeris, mucronem anteriorem in faciem hostis porrexeris, corpus circumagas ita ut dexter prior sit, atque caput ictu / hastae quassato.

Si vero is vicissim eadem ratione utatur, hastam humo infige ex latere tuo dextro, atque obviam eis si processeris, / intra manum tuam utramque ictum eius excipito, subito interea appete visum eius impulsu, et ab eo discedes in tuta tamen tuis / defensione.

C93, 185v

Ainn rechts unnd ain linncks leger.

Item schiktn dich Allso inn das leger von der rechten seiten / stannd mit deinnem rechten fuoß vor und hald dein stanngen an deinner Rechten seitten in der sterckn / stat er dann allso gögen dir von seinner lincken seitten auch inn der sterckn sein lincten fuoß furgeseßt / und hapt baid ain annder angepunden so thüo sain wellestü im zü seinem gesicht stossen indes wechsel / im durch von seiner rechten aüf sein lincken seitten volg mit deinnem lincken schennckel hinach / und stoß im zu der Brust verseßt er dir das so volg mit deinnem rechten schenckel hinnach und wind / im deinen hindern Orth zwischen seinnen baiden Armen zü seinnem angsict binnein.

Windt er / dir also nach deinnem gesicht so trit mit deinnem rechten füoß hinnein und seß im das ab mit deinnem / hindern orth Indes wechsel im mitt deinnem Orth dürch und stoß im auch nach seinem angesicht / seßt er dir das ab so ubergee unnd schlag im mit ainnem haw nach seinem haupt.

Schlect er / dir allso nach deinnem haüpt so seß dein stanngen aüf dir erden aüf dein lincken seitten und gee im / mit deinner stanngen entgegen und verseßß im das zwifche deine baid hend Indes stich im nach seinem / gesicht und zeüch dich damit zü rückn inn gute versaßung.

A left and right guard.

Position yourself thusly for this guard on the right side: stand with your right foot forward and hold the staff at the strong, on your right side (as depicted by the left figure). If the adversary then stands against you in the strong on his left side, with his left foot leading, and you have both bound each other, then position yourself as if you want to thrust into his face. Instead, immediately change through from his right side to his left, pass forward with your left foot, and thrust into his chest. But if he sets your thrust aside, then immediately pass forward with the right foot, wind your back point between both of his arms, and thrust into his face.

If the enemy attempts this same thrust to your face, then step to the inside with your right foot, parry his thrust with your back point, simultaneously change him through with your long point, and thrust into his face. If he sets your thrust aside, then place your left foot in front, extend your front point into his face,[1] turn your body so that your right foot is leading, and strike his head.

On the other hand, if the adversary attempts this same strike to your head, then drive your staff toward the ground on your right side,[2] advance toward him, parry his strike between both of your hands, simultaneously thrust into his eyes, and withdraw from him in a strong defensive guard.

---

1. This technique is described in the German as *obergehen*, "overgoing," and appears to be a synonym for *uberlauffen*, "overrunning."
2. The Dresden text reads *lincten seitten*, "left side."

Shortstaff 7

Cod. Vinob. 10825, 157r

Impulsus quo visus adpetitur contra inferam aversionem per hastae mucronem utrumque.

Hanc in formam hoc pacto te accomodato, hastam in latere dextra contineas, manum sinistram hastae mediaetati adplicabis, / manum vero dextram iuxta pedem dextrum, atque mucro hastae anterior sit contra adversariis faciem porrectus. Verum si is in habitu / aversionis tibi contrarius perstiterit, sinistrum pedem praeponens, hastaque eius terrae de loco prioris infixa sit, manus vero eius / dextra recta iuxta posteriorem mucronem constituta, dextra pede insequutus, visum eius impulsu adpetes.

Sin vero hostis eadem / ratione te adgressus fuerit, dextro pede intro procedes, atque in ipso incessu recta sub oculos eius hasta tua procedito, et impulsum / adversariis avertito, interea autem inflectes tuae hastae mucronem posteriorem intra eius brachium utrumque versus collum propellito.

Caeterum si habitu consimilis contra te is fuerit usus, dextro pede recedito, atque punctionem eius mucrone anterioris eludes, nec / non ex hastarum contactus habitu caput hostis quassabis. Interea autem ex tuta defensione recede ab adversario, attamen / vulorum curam diligentem habebis, ne instando te sequatur de improviso.

C93, 186r

Ein gesicht stoß gögen ainem underen abnemen.

Item schickn dich Allso mit dem zufecten Inn den gsicht stoß / halt dein stanngen aüf deinner lincken seitten mit deiner lincken hand im der halben stanngen dein / rechten hand hindern bey deinnem recten schennckel deinen vordern orth gögen seinem gesicht Stat / er dann inn dem abnemen gögen dir seinnen lincken fuoß furgeseßt sein stangen auf der Erden die / recht hand gestract hinden bey dem orth so volg mit deinnem Rechten schenckel hinnach und stoß im / nach seinem gesict.

Stost er dir also nach deinnem gesicht so trit mit deinnem recten schenckel / hinnein unnd gee mit dem trit mit der stanngen krad fur sein angsicht und seß im seinen stoß ab / indes wind im deinnen hindern orth zwischen seinen baiden armen hinnein zü seinem hals.

Winde er dir allso nach deinnem hals so seß deinnen rechten schenckel widerumb zü rüctn und seß / indas ab mit deinnem vordern orth und schlag Im auf dem annpund zü seinem kopff zeüch dich / damit zü rüctn In gütter versaßung und hab dein es gesicts güt ache das er dir mit nachraise.

A thrust to the face against a lower *abnemen*.[1]

To execute a face-thrust, position yourself as follows: hold the staff on your right side[2] with your left hand in the middle, your right hand back by the right leg, and your forward point extended toward the opponent's face (as depicted by the left figure). If he stands against you in a low guard with his left foot leading, his staff pointing at the ground, and his right hand by the back point (as depicted by the right figure), then pass forward with your right foot and thrust into his eyes.

If he thrusts toward your face in this same way, then step to the inside with your right foot, strike toward his face, set his thrust aside, simultaneously wind your back point between his arms, and thrust into his throat.

If he attempts a similar attack to your throat, then step back with the right foot, set his thrust aside with your front point, and strike his head from the bind. Then withdraw from him in a strong defensive guard, and make sure that he does not suddenly pursue you.

---

1. German, "taking off." See Latin *habitus aversionis*, "parry; setting-aside."
2. The Dresden edition erroneously reads *lincten seitten*, "left side."

Shortstaff 8

Cod. Vinob. 10825, 157v

Duae aversiones ex latere utroque.

Quum rite hoc habitu iam praedicto uti voles, tuum dextrum pedem praefigere non dubitabis, porrecto brachio hastam terrae inclinato ab / parte eius priori, manus sinistra in hastae medio constituta sit, dextra vero superne iuxta faciem tuam. Et si adversarius modo / aequali contra te constiterit de latere suo sinistro, pede laevo intro procedes, atque mucronem supernum intra eius brachium utrumque ad visum / inflectito. Sin autem is eum conatum averterit, dextro pede confestim insequutus, mucrone hastae tuae posteriori ex habitu aversionis / brachium dextrum conquassabis.

Verum si hostis idem fuerit molitus, te sinistrum praeponente pedem, hostis impetum mucrone / anteriori de manu tua dextra avertes, celeriter autem pede insequutus dextro, pectori eius mucronem posteriorem hastae / infigito. Caeterum si is eum exceperit conatum, sinistro consequitor pede intro, atque dupliciter mucronem hastae utrumque contra eius / faciem propellito.

Si vero hostis eadem impetus ratione fuerit usus, tum dextrum referes pedem, et avertere mucrone anteriori / sedulo curabis adversariis impetum, interim vero mucronem hastae posteriorem versus visum eius propellito vel / pectus, et ab eo discedes in tuta tuis defensione.

C93, 186v

Zwaii absen vonn baiden seitten.

Item schickn dich allso Inn diseß absenen mit dem / zufecten stee mit deinnem rechten fuoß vor dein stanngen aüf der erden mit gestractiem / arm dein lincke hannd inn mitte der stanngen die rect oben vor deinnem gesict stee er auch / allso gleich gögen dir von seinner lincken seitten so trit mit deinem lincken schenckel hinnein / und wind im deinnen obern orth zwischen seinen Armen zü seinem gesicth verseßt er dir das / so volg mit deinem rechten fuoß hinnach und haw im aüß dem absßen deinen hindern orth / zu seinem rechten arm.

Schlect er dir allso nach deinnem rechten arm und du mit deinem linckn / fuß vorsteest so nimb im das ab mit deinnem vordern ort in der rechten hand indes volg mit deinem / rechten schenckel hinnach und stoß im deinen hindern ort zß seiner prßst verseßt er dir das / so trit mit deinem lincken schennckel hinein und wind im zwifach mit deinem ortern zu / seinem gesict.

Windt er allso zwifach ein so zeüch deinen rechten schenckel zü rückn und seß im / das ab mit deinem vordern orth Indes stoß im mit deinem hindern orth zü seinem gesicht oder der / prüst und zeüch dich damit zü rückn in güte versaßung.

Two parries from both sides.

For the aforementioned parry, position yourself thusly: stand with your right foot forward, your arms extended, your front point touching the ground, your left hand in the middle of the staff, and your right hand in front of your face (as illustrated by the left figure). If the adversary stands against you in the same position on his left side (as illustrated by the right figure), then step to the inside with your left foot and wind your upper point between both his arms and into his eyes. However, if he parries you, then immediately pass forward with your right foot and strike his right arm with your back point.

On the other hand, if he strikes your right arm in this same manner when you are standing with your left foot forward, then set him aside with the front point by your right hand, immediately pass forward with the right foot, and thrust your back point into his chest. If he parries your thrust, then follow to the inside with your left foot, and twice wind into his face with both points.

If he attacks you with the same wind, then bring your right foot back, set him aside with your front point, simultaneously thrust your back point into his eyes or chest, and withdraw from him in a strong guard.

SHORTSTAFF 9

Cod. Vinob. 10825, 158r

Habitus, quo visus adpetitur contra formam aversionis.

Ad eum habitum hasta in congressu regenda in hunc modum te compones, sinistrum praefiges pedem, hastam in manu dextra de femore / itidem dextro contineas, mucro autem versus adversariis faciem sit porrectus. Sin vero forte evenerit, ut is in habitu aversionis contra / te constiterit dextrum pedem praefigens, hastamque in medio utraque manu recta pro facie contineat, tum pede dextro intro concedes, et mucronem / hastae posteriorem in faciem eius propellito de visus latere dextro. Verum si is averterit confestim anteriorem iniscito in eius pectus / interea autem dextro pede reducto, dextrum latus adversariis hastae uti pulsato.

Sed si contra te eodem habitu uti conatus est, / sinistro ad eum concedas, atque impetum eius intra tuas manibus ex latere tuo dextro avertere curabis, celeriter autem hastae / partem anteriorem in visum hostis propellito. Quum vero idem simili ratione exceperit, pede dextro insequutus, adversariis pudenda / mucrone posteriori adpetes.

Sed si is eadem modo te inferne adgreditur, sinistro pede retrorsum concedes, nec non eius impulsum / mucrone posteriori excipies, statim vero pedem sinistrum rursus intropones, et duplici impulsu visum eius adpetito, tuto interim / ex ea habitu ab adversario recede,

C93, 187r

Ainn gsicht stoss gögen ainnem absetzen.

Item schickn dich allso Mit dem gesicht stoß inn / dem zufechten stee mit deinnem lincken fuoß vor dein stanngen in deiner rechten hand auf deiner / rechten hufft den orth gögen seinnem gesicht stet er dann gögen dir inn dem abseßen seinen recten / fooß furgeseßt sein stanngen inn der hutt mit seinen baiden henden gestrackt vor seinem angsicht / so trit mit deinem rechten fuoß hinnein und wind im deinen hindern ort zu seinem gesicht seiner / rechten seitten sesst er dir das ab so wind im mit deinem andern Ort zü seiner prüst indes seß deinen / rechten schenckel zü rückn und schlag im mit deinner stangen nach seiner rechten seitten.

Schlect er dir allso nach deiner rechten seitten so trit mit deinnem lincten schenckel hinnein und / seß im das ab ~~mit~~[1] zwischen deinen baiden henden auf dein lincken seitten indes stoß im mit / deinem vordern orth zu seinem gesicht verseßt er dir das so volg mit deinem rechten fuoß hinnach / und stoß in mit deinnem hindern orth zü seinnen gemechten.

Stost er dir also unden zu so trit mit / deinem lincken fuoß hinder sich und seß im das ab mit deinem hindern orth indes trit mit deinem / lincken fuoß hinnein und stoß im mit ainem zwifachen stoß nach seinem gesicht trit damit / zü rückn inn guter versaßung.

A thrust to the face against a parry.

For this technique, position yourself thusly during the approach: stand with your left foot forward, hold the staff with your right hand near your right hip, and extend your forward point toward the adversary's face (as illustrated by the left figure). If he parries you with his right foot in front and both hands extended in front of his face, holding the middle of his staff (as illustrated by the right figure), then step to the inside with your right foot, wind him, and thrust your rear point into his face on the right side. But if he sets you aside, then immediately wind him and thrust your forward point into his chest; at the same time, bring your right foot back and strike into his right flank.

On the other hand, if the enemy strikes at your right side in the same manner, then step toward him with your left foot, parry his attack between both hands on your left side,[2] and simultaneously thrust your forward point into his eyes. If he parries your thrust, then pass forward with your right foot and thrust your rear point into his groin.

But if the adversary attacks your groin with a similar thrust, then step back with the left foot, set him aside with your rear point, immediately step to the inside with your left foot, thrust twice into his face, and withdraw from him in a strong guard.

---

1. The word *mit* appears to be crossed out in the Dresden edition and is entirely absent from the Vienna MS, which reads *und sesz im das ab zwischen deinem baiden henden*.

2. The Vienna reads *ex latere tuo dextro*, "from your right side."

Shortstaff 10

Cod. Vinob. 10825, 158v

Duae irruptiones unde habitus hostis prosternendi formatur.

Quum in adversariis conspectum hasta athletice regenda concesseris, sinistro pede intro procedes, atque hastam vultus infligito hostis / de latere tuo dextro. Verum si adversarius id exceperit, dextrum pedem reducito, atque hastam voluendo per manus si retraxeris mucrone / longo caput eius quassato.

Sin vero eodem habitu hostis utatur, tum intra manus tuas hasta excipies ictum eius. Interea / autem intro procedito pede dextro, atque mucronem hastae anteriorem versus visum eius impellito. Avertente id adversario, rotando / circumflectes tuam per eius hastam, insequutus sinistro pede, dextrum hostis latus mucrone posterioris pungito.

Idem vero si is conabitur, / impulsum excipies, et contra adversarium pede dextro si processeris, mucrone posteriori faciem eius adpetito.

Si autem / consimili ratione te adgressus fuerit, impetum hostis excipere non cessabis, dextroque introgreditor ad hostem, atque in habitu aversionis / mucronem hastae tuae anteriorem collo eius adiungito, et dextrum pedem pedis sinistri adversariis popliti si inieceris, superne / propuleris, inferne autem adtraxeris, adversarium prosternes.

C93, 187v

Zway ein brechen darauß ein wurff geet.

Item wann du mit dem zufechten zu dem mann kompst so / trit mit deinem lincken schenckel hinein und stoß im mit deinner stanngen nach seinem gesicht von / deinner rechten seitten verseßt er dir das so seß deinnen rechten schennckel zü rückn und laß dein stangen / durch dein hand schiessen und schlag inn mit dem langen ort zü seinnem haüpt.

Schlect er dir allso nach / deinnem haüpt so verseß im das zwischen deinnen henden in dein stanngen indes trit mit deinnem rechten / schennckel hinnein und stoß im mit deinnem vordern orth nach seinem gesicht seßt er dir das ab so wechsel / durch an seinner stanngen volg mit deinem lincken fuoß hinnach und stoß im mit deinem hindern / orth nach seinner rechten seitten.

Stost er dir allso nach deiner rechten seitten so verseß im das und trit mit / deinnem rechten schennckel zü im hinnein Indes stoß im mit deinem hindern Ort zü seinem gesicht.

Stost er dir also nach deinem gesicht so verseß im das trit mit deinem rechten schenckel zü im hinnein / und in dem abseßen falt im mit deinnem vordern orth an seinen hals und mit deinem rechten fuoß hinder / seinnen lincken inn den hacten trüctn oben von dir und zeüch unden zü dir so fesst er zü rückn.

Shortstaff 10

Two break-ins followed by a throw.

When you approach the adversary, step to the inside with your left foot and thrust into his face from your right side. But if he parries your thrust, then bring your right foot back, draw the staff back through your hands, and strike his head with the long point.

On the other hand, if the enemy attempts the same strike to your head, then parry his blow between your hands on the staff. In the same instant, step to the inside with your right foot and thrust your front point into his eyes. If the adversary parries your thrust, then change him through by winding around his staff, pass forward with your left foot, and thrust your back point into his right flank.

But if he attempts this same thrust to your right side, then parry his blow, step toward him with your right foot, and thrust your back point into his face.

If he attacks you with a similar thrust to the face, then parry him, step toward his inside with the right, and as you parry, hook your front point around his throat, place your right foot behind his left knee (as depicted in the illustration), push up, pull him down toward you, and throw him onto his back.

Note that a nearly identical throw is described in Halberd 14, but the position of the weapons is different. This may be an artistic oversight.

Shortstaff 11

Cod. Vinob. 10825, 159r

Duo impulsus versus supernas nuditates ipsorum.

Si hasta more athletico regenda ad hostem processeris, pedem sinistrum praefiges, directe brachia tua in hasta contineas, manu / dextra super capite tuo constituta, mucro autem contra adversariis pectus sit porrectus. Et si is vicissim in impulsu superno / contra te constiterit, sinistrum praeponens, dextro pede confestim insequitor, hastam rotando manu dextra versus latus tuum dextrum / mucronem longum ex ipsa rotatione hastae in faciem eius infigito. Cum vero impulsum animadverterit, atque eum eluserit se defendendo, / mutando de eius latere sinistro in latus dextrum si transieris hasta, pectus adversariis punctione adgreditor.

Et si is eadem / ratione utetur, impetum eius avertes, et dextro pede consequitor, interea mucronem hastae tuae posteriorem in faciem hostis / animose impellito, nec non in ipso inflexionis habitu pedem dextrum si reduxeris retrorsum, mucronem longum pectori ipsius infigito. Sin autem id exceperit, rursus dextro intro concedes, et anterioris mucrone caput feriundo quassabis adversariis.

Verum eum is idem fuerit molitus, id avertes hastam circumflectendo, interim vero dupliciter faciem pungendo adgreditor, / atque in tuta tuis defensione ab adversario recedas.

C93, 188r

Zwen stoß zu fur oberen plossen.

Item wann Du mit dem zufecten zü dem mann / kumpst so trit mit deinnem lincken schennckel hinnein und halt deine arm gestrackt inn / der stanngen dein rechte hannd ob deinnem haüpt den orth zü seinner prüst stat er dan auch / allso gögen dir inn dem obern stoß zü deinner obern plose seinen lincken fuoß furgefeßt so volg / mit deinnem rechten schennckel hinnach und wind mit deiner rechten aüf dein rechten seiten / und inn dem winden stoß im dein langen orth In sein gesicth wirt er des stoß gewar und verseßt / dir den so wechsel durch von seinner lincken aüf sein rechten seiten und stoß im zü seinner / prüst.

Stost er dir also zü so seß im das ab und volg mit deinnem rechten schenckel hinnach indes / wind im deinnen hinder orth zü seinnem gesicht und in deinen winden so trit mit deinem / rechten fuoß zü rückn unnd stoß im deinnen lanngen Orth zü seiner prüst verseßt er dir das / so trit mit deinnen rechten fuoß wider hinnein und haw im mit dem langen orth ober zü seinnem / haüpt

Schlecht er dir allso oben zß so seß im das ab mit ainnem winden an deiner stanngen / Indes stoß im zwifach nach seinnem gesicht unnd trit damit zü rückn in güter versaßung.

Two thrusts to his upper openings.

When you approach the enemy, set your left foot forward and hold the staff with your arms extended, your right hand above your head, and your point aimed at his chest (as illustrated by the left figure). If he in turn stands against you in the same position with his left foot leading (as illustrated by the right figure), then immediately pass forward with your right foot, wind the staff onto your right side with your right hand, and thrust your long point into his face. If he anticipates your thrust and parries, then change the staff through from his left side to his right side and thrust into his chest.

If he thrusts toward you in this same way, then set him aside, pass forward with your right foot while winding your back point hard into his face, bring the right foot back during the wind, and thrust your long point into his chest. If he parries you, then step to the inside again with your right foot and strike your front point into his head.

If he attempts this same strike to your head, then set him aside by winding, simultaneously thrust into his face twice, and withdraw from him in a strong guard.

SHORTSTAFF 12

Cod. Vinob. 10825, 159v

Mucro longus contra hastarum collisionem primam.

In accessu contra hostem ex eo ictu, qui ab aure vibratur contra eum, ferire memineris dupliciter, et si propius ad eum processeris, / sinistrum pedem praeponito, manum dextram femori adplicabis dextro, mucro hastae contra hostis faciem sit porrectus, confestim / vero insequutus pede dextro, eius collum mucrone longo figes.

Sin autem is eadem ratione te adgressus fuerit, eum sinistrum / tu praefixeris, atque in habitu hastarum collisione constiteris, mucrone anterioris impetum hostis impedito, et insequutus dextro / pede, caput eius ictu adpetes. Verum in ipso ictu, collo ex parte anterioris hostis circumflectendo hastam adplicabis, postea eius / pedi dextro laevum pedem anteponito, superne propelles, et diligenter curabis, ut adversarium hoc habitu infirmum reddas.

Caeterum / si is idem moliatur, insequitor pede laevo, atque eius mucronem hasta tua eludes, interimque mucronem anteriorem versus faciem hostis / impellito. Eo autem impulsum a te accurate vibratum avertente, hastam circumflectes, et rursum faciem eius adpetito.

Verum / si is gemino impulsu te fuerit adgressus, mucrone anteriori eum avertes, et mucronem posteriorem pectori eius infigito, ex eo / itaque habitu retrocedes in tuta tuis defensione.

C93, 188v

Ein lannger Orth gögen ainem anpünd.

Item wann du mit dem Zufecten zü dem mann geest / so haw dich aüß dem sturtzhäw freii zwifach zü im hinnein so du dann fur den man kumpst so stee / mit deinnem lincken fuoß vor dein rechte hannd aüf deinner rechten huff dein stanngen mit deinem / orth gögen seinnem gesicht Indes volg mit deinnem rechten schenckel hinnach und stoß im deinem / lanngen orth zü seinnem hals.

Stost er dir allso zü deinnem hals und dü mit deinnem lincken / fuoß vorsteest in dem anpünd gögen im so nimb im das hinweckn mit deinnem vordern orth Indes / volg mit deinnem rechten fuoß hinnach und schlag in nach seinnem haüpt und in dem schlagen so / wind Im dein stanngen vornen umb seinnen hals trit mit deinnem lincken fuoß fur seinen rechten / und truckn oben von dir und shaw ob dü im mochtest die schwech abgewinnen.

Begert er dich allso zü / swechen so volg mit deinnem lincken fuoss hinnach und nimb im seinen orth mit deinner stangen / hinweckn Indes wind im deinnen vordern ort zü seinnem gesicht verseßt er dir das so wechsel im durch / und stoß im widerumb aüf die vorgeschribnen stat zü dem gesicht.

Stost er dir also zwifach nach dem / gesicht so seß im das ab mit deinnen vordern orth und wind im denn hindern ort zü seiner prüst zeüch / dich damit zü rückn inn güte versaßung.

The long point against a bind.

When you approach the opponent, strike a double *sturtzhaw*,[1] move closer to him, place your left foot forward, bring your right hand to your right hip, and extend your point toward his face (as depicted by the left figure), then immediately pass forward with your right foot and thrust your long point into his throat.

If he thrusts thusly at your throat while you are standing in the bind with your left foot forward (as depicted by the right figure), then parry him with your front point, pass forward with your right foot, and strike his head; then, wind your front point around his neck, set your left foot in front of his right foot, and push forward from above so that you weaken him.

If he tries to weaken you, then pass forward with your left foot and parry him while winding your forward point into his face. If he sets your thrust aside, then change him through and thrust into his face a second time.

If he thrusts twice at your face in this way, then set him aside with your front point, wind your rear point into his chest, and withdraw from him in a strong guard.

---

1. German, "plunge strike."

SHORTSTAFF 13

Cod. Vinob. 10825, 160r

Ictus conversus contra aversionem.

Hoc modo te compones in predictum habitum, regendo ictum geminum de pectore formatum versus adversarium porrecte ferias, et si prope / ad ipsum perveneris, contra hostem dextro pede consistas praefixo. Verum celeriter hastam rotando, manu conversa hoc ita fiat, ut hasta / manu inversa retrorsum supra humerum ita voluatur, ut mucro anterior sit, caput eius conquassato, et rursum manu sinistra / hastam si adpraehenderis, mucronem anteriorem in visum eius porrigito.

Sin autem versus adversarium sinistro praeposito constiteris porrectus / manibus hastam contineris, mucroque versus terram vergat, tum eius ictum et punctionem avertes hasta intra manum utraque / introgressus pede dextro, mucronem brevem iniscias manu dextra contra eius faciem. Id si exceperit, dextro reducto, vultus hostis / mucronem longum impingito.

Ceterum adversario eodem habitu adversus te utente dupliciter, utrumque impulsum mucrone anteriori / avertes, confestim vero insequutus pede sinistro, mucronem posteriorem in faciem impellito adversariis dupliciter. Et si hostis eum conatum / animadverterit, atque excusserit, quantum poteris, hasta dextrum latus eius mucrone longiori quassabis. Sin autem rursum exceperit / ictum, tum dupliciter contra pectus eius hastam iniscito, atque ab adversario tuta recedas.

C93, 189r

Ainn verkerter schlag gögen ainem abnemen.

Item schickn dich allso Mit disem zufechten haw dich freylang / mit ainnem zwifachen prüsthaw zü im hinein wann du dann zü im kompst so trit mit deinnem rechten / fuoß zü im hinein Indes laß die stanngen behennd uberlauffen und schlag im nach seinem haüpt mit verkerter / hannd greif mit deinner lincken hannd winderumb inn die stanngen und wind Im deinnen vordern orth / zü seinnem gesicht.

Stastu dann allso gögen im mit deinnen lincken fuoß vor mit gestrackten armen Inn deiner / stannger den ort gögen der Erden so nimb im seinnen haw und stich ab zwischen deinen baiden henden / In dein stanngen trit mit deinem rechten schennckel hinnein und wind im deinnen kurßen / orth in deiner rechten / hannd zü seinem gesicht verseßt er dir das so trit mit deinnem rechten füoß widerumb zü rückn und stoß / Im deinnen stanngen orth zü seinnem gesicht.

Stost er dir allso zwifach nach deinem gesicht so nimb im das hinweckn / mit deinnem vordern orth Indes trit mit deinnem lincken füoß hinein und stoß Im deinen hindern orth / zwifach zü seinnem gesicht wirt er des zwifachen stoß gewar und verseßt dir den so schlag im behend mit deinnem / lanngen orth nach seiner rechten seitten verseßt er dir das furbas so wind im zwifach nach seiner prüst / trit damit zü rückn inn güte versaßung.

An inverted strike against an *abnemen*.[1]

Position yourself thusly for this technique: attack the adversary with a double *prusthaw*[2] and, as you close with him, set your right foot forward. At that moment, strike his head by swinging your staff around with an inverted hand (as illustrated by the left figure), seize the staff again with your left hand, and thrust your front point into his eyes.

On the other hand, if you are standing against him with your left foot forward, your arms extended, and your long point toward the ground (as illustrated by the right figure), then parry his strike and his thrust between your hands on the staff, step to the inside with your right foot, and wind your short point into his face with the right hand. If he parries you, then bring your right foot back and thrust your long point into his face.

If he thrusts toward your face in the same way, then set both of his thrusts aside with your front point, simultaneously pass forward with your left foot, and thrust your back point into his face twice. If he anticipates your double-thrust and parries, then quickly strike his right side with your long point. However, if he parries your strike a second time, then wind him, thrust twice into his chest, and withdraw in a strong guard.

---

1. German, "taking off." See Latin *habitus aversionis*, "parry; setting-aside."
2. German, "breast strike."

SHORTSTAFF 14

Cod. Vinob. 10825, 160v

Impulsus per quem brachium adversariis debilitatur contra impulsum, quo pudendo adpetuntur.

Quum in conspectum adversariis concesseris hasta rite regenda, pede sinistro procedes, et hastam supra caput manu dextra contineas / manum vero sinistram infra medium hastae adplicabis, atque eo modo hostis cubitum sinistrum pungito, hoc itaque habitu brachium eius / infirmabis.

Sed si is idem fuerit contra te molitus. Sique hastam dextra manu super capite continueris, et sinistrum pedem praefixeris, / hasta sursum levanda si averteris pudenda eius, mucrone anterioris adgreditor, et eo impulsu eius conatum irritum facies. Inde pede dextro intro procedes, atque mucronem hastulae posteriorem versus adversariis visum inflectendo impellas.

Caeterum si eadem / impulsus ratione is utiis conatus est, mucrone anterioris eum avertes, et si hastam recta ante faciem tuam continueris, dextro introgredere / contra hostem, et hastam adversariis ex manu eius anterioris excutias mucrone tuo posterioris, atque mucrone anterioris / faciem ipsius adpetito. Sin vero id hostis exceperit, dextrum pedem reducito retrorsum, et mucronem hastae longiorem, pectori / eius infigito, id itaque si perfeceris, ab adversario tuto recedas.

C93, 189v

Ein gewicht stoß gögen ainem gemecht stoß.

Item wann du mit dem Zufechten zu dem mann komst / so trit mit deinnem lincken schennckel hinnein und halt dein stanngen mit deinner rechten hand / ob deinnem haüpt dein lincken hand wol davornen Inn deiner stanngen und stich im nach seinem / lincken Ellenpogen so nimbstu Im das gewicht.

Stost er dir allso nach deinnem gewicht deinner linken / arms und dü dein stanngen auch Inn deinner rechten hannd halt ob deinnem haüpt dein lincken / fuoß furgseßt so seß im das ab ubersich auf an deiner stanngen Indes stoß im nach seinen gemechten / so trifftü In mit deinnem stoß unnd der sein ist umb sünst Inn dem volg mit deinnem rechten schenckel / hinnein und wind Im deinnen hindern orth zü seinem gesicht.

Windt er dir also zü deinem gesicht so / seß im das ab mit deinnem vordern orth und halt dein stanngen trad vor deinem angsicht in dem / trit mit deinnem rechten fuoß zü Im hinnein und schlag Im sein stanngen mit deinnem hindern orth / auß seinner vordern hannd und stoß Im mit deinnem vordern orth zü seinem gesicht seßt er dir das ab so / zeüch deinen rechten schennckel zü rückn und scheuß im dein lanngen orth zü seinner prüst trit damit / zü rückn inn güter versaßung.

A thrust to the elbow against a thrust to the groin.

As you approach the adversary, step forward with your left leg, hold the staff with your right hand above your head and your left hand lowered, gripping the middle of the weapon, and thrust toward his left elbow to pin his arm (as depicted by the left figure).

If he tries to thrust at your left elbow while you are holding your staff with the right hand above your head and your left foot leading (as depicted by the right figure), then set him aside by lifting your staff up high and thrust your front point into his groin, thus incapacitating him. From there, advance to the inside with your right foot, wind him with your back point, and thrust into his eyes.

If he attempts a similar thrust to your eyes, then set him aside with your front point, hold the staff directly in front of your face, step toward him with your right foot, strip the staff out of his leading hand with your back point, and thrust your front point into his face. If he sets you aside, then bring your right foot back, thrust your long point into his chest, and withdraw from him in a strong guard.

Shortstaff 15

Cod. Vinob. 10825, 161r

Propugnaculum seu castrum ex libra contra impulsum impetuosum.

Libra multifarius sunt habitus. Primus est, eum mucronem posteriorem pro facie contineas manu dextra, anterior terrae sit infixus / manus sinistra medium hastae, et si posteriorem in latus dextrum inflexeris. Secundus ex primo reddetur. In id igitur genus propugnaculis / ita te adaptabis, tibiis paribus insistes, hasta humum contingat, manum sinistrum hastae medietati adplicabis, dextra vero in / latere tuo dextro sit collocata, atque corporis habitu in libram te compones. Interim vero pede laevo intro procedas, et longum mucronem / contra pectus adversariis impellito.

Verum si hostis idem contra te usurparit, pedemque sinistrum tu praeposueris, hastam pedi sinistro / adplicaris ita, ut mucro eius terram contingat, manus vero dextra supra caput iuxta alterum hastae mucronem sublata, tum dextro / [crossed out] ingreditor et mucrone anterioris ipsius conatum eludito manu dextra, subito autem retrorsum pede dextro concedes, atque / ex totis viribus impetuoso impulsu contra eius vultum utitor mucrone longo porrecto.

Caeterum si idem adversus te hostis moliatur, / pede sinistro in triangulum si concesseris, eius impulsum devitabis, celeriter autem dextro intro procedes pede, rotata hasta caput / eius quassabis, interimque manu sinistra rursus hastae adplicata, ab adversario dupliciter tuto deflectes.

C93, 190r

Ein leger in der wag gögen ainem gewalt stoß.

Item schickn allso mit Dem zufechten inn disem / leger stee mit gleichen fuoßen zu samen dein stanngen aüf der Erden die linck hannd miten inn der / stanngen die recht Inn deinner rechten seitten unnd gib dich mit deinnem leib Inn die wag Indes trit mit / deinnem lincken schennckel hinnein und stoß Im deinen lanngen orth nach seiner prüst.

Stost er dir also nach / deinner Brüst unnd dü mit deinnem lincken fuoß vor steest dein stanngen aüf deinem lincken schenckel / aüf der Erden dein rechten hannd ob deinnem haüpt bey deinem orth so trit mit deinem rechten schenckel / hinnein und seß im das ab mit deinem vordern orth deinner rechten hannd In dem trit mit deinnem rechten / fuoß wider zü rückn unnd stoß Im mit dem gwalt stoß mit deiner sterckn zu seinem gesicht mit deinem langen / ort.

Stost er dir allso mit der sterckn zü deinnem gesicht so trit mit deinnem lincken fuoß inn triangel / so gestü Im auß seinnem stoß Indes trit mit deinnem Rechten schennckel zü Im hinein und laß dein stangen / uberlaüffen und schlag Im nach seinnem haüpt Inn dem greiff mit deinner lincken hand widerumb inn / dein stanngen und wend dich damit zwifach von im zü rückn Inn güte versaßung.

Spear and Shortstaff

A guard in the balance against a forceful thrust.

There are several guards from the balance. First, hold the back point in front of your face with your right hand, your front point touching the ground, and your left hand in the middle of the staff, then wind the back point onto your right side.[1]

The second is in response to the first. As you approach the opponent, assume a balanced stance with your left foot forward, your long point touching the ground, your left hand in the middle of the staff, and your right hand holding the back point in front of your face, on your right side (as depicted by the right figure); then, immediately step to the inside with your left foot and thrust your long point into his chest.

If he thrusts thusly at your chest when you are in a balanced stance with your left foot forward, your staff touching your left foot in such a way that its long point is on the ground, and your right hand holding the back point above your head, then immediately advance with your right foot, set him aside with your front point using your right hand, bring your right foot back, and thrust the long point into his face with all your might.

If he tries to do the same thing to you, then step in a triangular motion with your left foot, void his thrust, immediately step to the inside with the right, release your left hand, strike his head by overrunning him, grip the staff again with the left hand, and, by side-stepping twice, withdraw from him in a strong guard.

---

1. Note that there is no corresponding German text for this sentence.

Shortstaff 16

Cod. Vinob. 10825, 161v

Habitus longi mucronis cum aversione contra mutatorium cancellatum.

Quum prope ad hostem accesseris hastam athletice gubernando dextrum pedem intropones, hasta terrae sit proclinata anterius, manum / dextram eius medietati adplicabis, sinistram vero iuxta mucronem posteriorem supra coxam sinistrum locabis. Sin autem adversarius / contra te in habitu cancellati mutatoriis dextrum pedem praeponens constiterit, mucrone sublato longiori, atque insequutus sinistro, visum / eius impulsu adpetito. Eum si hostis eluserit, dextro consequutus, mucronem posteriorem versus faciem adversariis vel pectus diriges.

Quum vero is gemino impetu contra te fuerit usus tuo exemplo provocatus, hasta sublata mucrone anterioris avertes ipsius impulsum. / Insuper pede sinistro insequutus, ex mutatorio cancellato mucronem posteriorem eius vultus impelles. Verum si hostis / tuum impulsum averterit se acriter defendendo, quantum poteris, latus eius sinistrum inflectendo dupliciter adpetes.

Sin vero / tibi eo pacto institerit adversarius, hasta media eius impetum avertes, insequutus confestim sinistro, mucronem posteriorem / versus latus adversariis sinistrum coniscito, verum ex eo habitu pedem sinistrum reducito, et caput eius animose mucrone / anterioris quassato, hoc itaque perfecto ab hoste recedes.

C93, 190v

Ein langer ort mit ainem abnemen gögen ainem geschrenckten wechsel.

Item wann du mit Dem zufecten zu dem mann komst, / so trit mit deinem rechten fuoß hinein, und halt dein stanngen vornen auff der erden, dein rechten hand / mitten dar im die linnckn hinden bey deinnem orth Inn deiner lincken hüff stat er dann allso gögen dir / In dem geschrennckten wechsel seinnen rechten fuoß furgeseßt so gee aüf mit deinnem lanngen orth / Indes volg mit deinnem lincken schennckel hinnach und stoß im nach seinnem gesicht seßt er der das / ab so volg mit deinnem rechten füoß hinnach und wind im deinnen hindern orth zß seinnem gesicht / oder der prüst.

Stost er die allso zwifach zü deinnem gesicht so gee aüf mit deinner stanngen und seß / Im das ab mit deinnem vordern orth volg mit deinnem lincken schennckel hinnach und stoß im deinen / hindern orth zü seinnem gesicht auß dem geschrennckten wechsel verseßt er dir das so wind im behennd / wider zwifach ein zü seinner lincken seiten.

Raist er dir allso nach so seß im das ab mit deinner / halben stanngen volg mit deinnem lincken schenckel hinnach und stich im mit deinem hindern orth / zü seinner lincken seitten Indes zeüch deinen lincken schennckel zü rückn und schlag Im mit deinem / vordern orth zü seinnem haüpt trit damit zü rückn inn güter versaßung.

A parry with the long point against a crossed *wechsel*.[1]

As you approach the enemy, step to the inside with your right foot, place your right hand in the middle of the staff and your left hand near the back point, by your left hip, and tilt your front point toward the ground (as depicted by the left figure). If he stands against you in a crossed *wechsel* with his right foot leading (as depicted by the right figure), then raise your long point, pass forward with your left foot, and thrust into his eyes. If he sets you aside, then pass forward with your right foot, wind him, and thrust your back point into his face or chest.

On the other hand, if the opponent thrusts twice at your face in the same manner, then raise your staff from the crossed *wechsel*, set his thrust aside with the front point, pass forward with your left foot, and thrust the back point into his face. If he parries your thrust, then wind twice into his left flank.

If he attacks you with this same technique, then set him aside with the middle of your staff while passing forward with your left foot, thrust the back point into his left flank, quickly bring your left foot back, strike his head with the front point, and withdraw from him in a strong guard.

---

1. German, "changer." See Latin *mutatorius*, "change."

Shortstaff 17

Cod. Vinob. 10825, 162r

Duo superni impulsus contra pectus formati de latere utriusque sinistro.

In eum habitum hoc modo te compones, contra hostem pede sinistro procede, et hastam de latere tuo dextro versus pectus ex parte / eiusdem sinistra inter eius brachium utrumque iniscito.

Sin autem is vicissim contra te constiterit in habitu impulsus supernis de latere / suo laevo, pedem sinistrum praeponens, pectusque tuum eodem modo adgreditur, tum ab hasta tua manum sinistram removebis, et hastam / adversariis iuxta mucronem eadem adpraehende, mucronem hastae tuae manu dextra inflectes sub eius axillam sinistram, contra adversarium / pede dextro procedes, et brachiis cancellatis si in latus dextrum deflexeris hasta utraque correpta, ita eum concludes, / ut prorsus nihil adversus te possit moliri.

Verum si te eadem ratione circundederit, hastam abiscere memineris, pedem sinistrum / exterius adversariis dextro postpones, manu dextra eius poplitem dextrum arripe, manum vero sinistram sub ipsius axillam / dextram circum corpus hostis regendo adplicabis, atque eo modo eum prosternes absque tuis corporis detrimento.

C93, 191r

Zweii ober stoß zü der Brust von fres linckenen seiten.

Item schickn dich allso mit disem zufechten trit mit / deinnem lincken fuoß zü im hinnein und stoß im mit deinner stanngen von deinner lincken / seitten zü seinner lincken prüst zwischen seinnen armen hinnein.

Stat er dann gögen dir auch / In dem obern stoss von seinner lincken seitten und hat seinnen linncken fuoß furgeseßt und stost / dir auch allso nach deinner prüst so laß dein lincke hannd von deiner stanngen und greif damit / In sein stanngen bey seinnem orth Indes wind im mit deiner rechten hannd deinnen Orth under / sein lincken Achsen unnd trit mit deinem rechten fuoß zü Im hinein wend dich damit aüf dein / rechten seitten mit Ererßweisen armen mit baiden stanngen so sperstü In das er zü kainer arbait / kumen kan.

Hat er dich allso beschlossen das dü mit deinner stanngen zü kainer arbait kumen / kanst so laß die stanng behennd falten seß deinnen lincken fuoß aüß wendig hinder seinen rechten / unnd greif mit deinner rechten hand nach seiner rechten kniepug und mit deiner lincken under sein / rechten Achsen wol umb dein leib hinnumb so wirfstü In das er dir kainer schaden kan züfugen.

Two thrusts to the chest from above, both on the left side.

Position yourself thusly for this technique: advance toward the enemy with your left foot and thrust the staff from your left side,[1] between both of his arms, and into his left pectoral.

If he in turn stands against you in the guard for a high thrust from his left side, his left foot leading (as depicted by the right figure), and he thrusts thusly at your chest, then release your left hand and seize his staff near the front point, wind your front point under his left armpit with the right hand (as depicted by the left figure), step toward him with your right foot, and turn onto your right side with your arms crossed, holding both staffs; you will trap him in such a way that he cannot do anything.

If he traps you with this same technique, then quickly let go of your staff, set your left foot behind his right foot,[2] seize his right knee with your right hand, reach your left hand under his right armpit and around his body, and throw him.

---

1. The Latin reads "from your right side," which conforms to neither the German text nor the title of this plate.
2. The German reads "set your right foot behind his left."

Shortstaff 18

Cod. Vinob. 10825, 162v

Prostratus qua ratione sit retinendus.

Hoc modo te accomodabis, contra adversarium ex ictu qui ab auribus utrisque vibratur, ferias, atque si propius accesseris, pedem / sinistrum si praefixeris, mucrone anterioris visum adversariis adpetito.

Sin autem idem contra te usurparit hostis, si et hasta athletice / regenda contra accedas, tum eius impetum inflexione ex loco hastae anterioris avertes, et pede dextro versus hostem procedito, nec non / mucrone longiori pectus eius adgreditor.

Caeterum si is eodem modo fuerit te adgressus, prope ad adversarium accede, atque impulsum / ipsius, hastae medietate intra manum tuam utramque eludito. Interim vero hastam supra caput reisciendo retrorsum habitu corporis in libram / te adaptabis, ilia adversariis manu utraque circumdabis, et hostem versum te convertes, cumque levaris eum sursum, infirmus ad rebellandum / factus est. Verum ex eo habitu prosternito hostem, et si fuerit prostratus, tum genu dextrum, ipsius pudendis intra eius pedem utrumque / adplicabis, sinistro adversariis sub dextrum devenies, et eius manum utramque vel guttur adpraehendes, et si fortiter undique deorsum depresseris, / eum facite captivum retinebus.

Sin autem id ab adversario tibi contigerit, diligenter in primis curabis, ne manus tuae captiventur. / Verum celeriter manu altera ita faciem adpraehendes, ut pollex mento illius sit suppositus, reliqui eiusdem manus digiti, pupillae adplicabis / acriter inhereat. Porro alteram manum eius pudendis impellito forti impetu, pedem qui liberior fuerit exere, sed celerrime reducto, / plurimum offendes eius pudenda ex eo habitu, atque his tribus habitibus adversarium repelles.

C93, 191v

Ainen geworffnen gefangen zuhallten das er nit aüf kan.

Item halt dich allso mit disem zúfechten haw dich in dem sturtzhaw / zü im hinein seß deinen lincken fuoß vor und stoß im mit deinem vordern orth zü seinem gesicht.

Stost er dir allso / zü deinem gesicht und dü auch inn dem zufechten bist gögen Im so seß im das ab mit ainem winden vornen an / deinner stanngen trit mit deinnem rechten fuoß zü im hinnein und stoß im mit deinnem lanngen ort zü / seinner prüst.

Stost er dir also nach deinner prüst so trit wol zü im hinein und seß im den stoß ab mitten inn / dein stanngen zwischen deinen baiden henden indes wirff dein stanngen uber deinnen kopff hinauß und gib / dich mit deinnem leib inn die wag und greiff mit baiden henden nach seinner waich trück in zü der und heb / in wol ubersich so hastu inn in der schwech indes wirff in unnder dich hastü in dann geworffen und zü fal / pracht so knie im mit deinnem rechten fuoß zwischen baid fuoß aüf seine gemecht und mit dem lincken / under seinner rechten und faß im baid hend oder die gürgel trückn faft allenthalben zür erden.

So behel stü / in bistü dann ir vergwaltiger und undergeworffen so hab erlendis güt acht das dir deine hend nit verschlossen / sonder für im gleich mit ainer hand inß angesicht den daümen underkin die andere finger / under den augapffel greiff allso starckn mit der anndern hand stoß in starckn zü den gemechten streckn / ainen fuoß der dir ain ledigsten ist und zeuch in geschwind wider an dich so gibstü im ein gütß zü den gemechten / mit disen dreien griffen zü ainem mal magstü in von dir bringen.

A throw from which he is restrained.

During the approach, position yourself for a *sturzhaw*,[1] move in closer, set your left foot forward, and thrust your front point into the opponent's face.

If he thrusts thusly at your face as you approach, then set his strike aside by winding with the forward part of your staff, step toward him with your right foot, and thrust your long point into his chest.

If he uses this same technique against you, then step toward him and parry his thrust with the middle of the staff, between your hands; in that instant, throw your staff backward over the top of your head, position yourself in a balanced stance, grab his groin with both hands, lift him up, and flip him over so that he cannot fight back. Once you have thrown him, put your right foot between his legs, knee him in the groin, hook your left leg under his right leg, seize either his hands (as depicted in the illustration) or his throat, and hold him down by putting pressure on every part of his body.

If he throws you down and tries to pin you, then make sure from the start that your hands are not captured, quickly seize his face with one hand (hook your thumb under his chin and press the other fingers into his eyes, gripping firmly), forcefully strike his groin with the other, kick out whichever leg is freer, and quickly draw it back, thus kneeing him in the groin. With these three techniques, you will defeat him.

---

1. German, "plunge strike."

SHORTSTAFF 19

Cod. Vinob. 10825, 163r

Media hasta ab utroque latere avertendi habitus cum impulsu.

    Ad hunc congressum hasta gubernanda athletice, hoc modo te compones. Si praefixo dextero pede steterit / adversarius utens contra te impulsum et appetit eo impulsu visum tuum: tum ita te para, dextrum tuum pedem / praefigito, et utroque bractico superius in altum protenso, hostem observes, et progreditor cum tuo laevo / pede celeriter contra eum, et mucronem tuum posteriorem impellito contra visum eius, hoc modo excipis / ipsius punctionem, et poterit uterque vestrum uti hoc habitum ad utrumque latus, vel tota vel media hasta, per / circum volutionem hastarum, in posteriori vel priori instantia.

Cod. Vinob. 10825, 163r

Ain halber stangen zue beider seitten ab zufechten stoss mit.

    Item halt dich also mit disem zufechten stat er mit ainem stoss also gegen dir mit seinem rechten / fuoss vor und stoss die nach dein gesicht so schick dich also das du mit deinem rechten fuoss vorsteest / und mit baidem aremen inn der hoch [...] und trit mit deinem linggen fuoss behend zue / im und wind in mit deinem hindern orth nach seiner gesicht so nimbst du im seiner spiess / ab und . . . diso . . . streick zue baiden seiten mit gannzen oder halben stanngen gebraucht . . . / . . . das winden der stangen im nach und vor . . . .

A half-staff parry to either side, with a thrust.

If the adversary takes position for a thrust during the approach, his right foot leading, and he thrusts at your face (as depicted by the left figure), then set your right foot forward, raise both arms up high (as depicted by the right figure), quickly step toward him with your left foot, and wind your back point into his eyes while parrying his thrust. You can use this technique on both sides, with either the middle of the weapon or the entire staff, by winding the staff in a circle and threatening him with the front or back point.

SHORTSTAFF 20

Cod. Vinob. 10825, 163v

Percussio, cum habitum aversionis.

Ita te para ad hunc habitum. Infligito ita longiore hasta tuam ictum superum, caput ipsius adpetenda, et consequere / tuo laevo pede.

Si hostis te petit eo modo percutiendi, arripe hastam ad medium ictum et impulsum egregie superius / in altum et progreditur cum tuo laevo pede contra hostem, excipe hoc modo avertendo ictum seu percussionem ipsius / media hasta. Et impellito posteriorem tuum mucronem intra sua duo brachia versa ipsius visum, et cum / anterioris mucrone, infligito hastam capiti eius, curam habendo in discedendo ab hoste, tuta tuis defensione, / ne reperiaris nudus.

Cod. Vinob. 10825, 163v

Ain schlag mit ainem absetzen.

Item halt dich also in disem stuck zufechten, schlag also mit lanngen stanng von oben ab zue / seinem haubt und trit mit deinem linggen fuoß hinnach.

Schlacht er also zue dir . . . die / stanngen zue halben stoß wol in die hoch und trit mit deinem linggen fuoß in auf / gegen und sez im damit in der miten der stangen sein streich ab und wind im mit / dem hindern orte zwischen saine baide armen zu dem gesicht und mit dem verndern ort schlag / im nach seinem haubt und yeuch von im ab mit guten verssassung.

A strike with a displacement.

Position yourself thusly for this technique: strike his head from above with your long point (as depicted by the left figure), and pass forward with your left foot.

If he strikes you in this way, then hold your halfstaff high in the air, advance toward him with your left foot (as depicted by the right figure), set his strike aside with the middle of your staff, wind your back point between both of his arms, thrust into his eyes, strike his head with your front point, and withdraw from him in a strong guard.

Shortstaff 21

Cod. Vinob. 10825, 164r

Inferior et superior hastarum contactus hastis mediis.

Si ita uterque vestrum utroque hasta contactum superiori et inferiori invicem contigeritis, tum laevo tuo pede progreditor, / et impellito eum posteriore mucrone versus visum ipsius: Si excipiendo hoc averterit, infligito ipsi / anteriorem mucronem capiti eius.

Si hoc modo ad te percutiet excipe hoc ei media et tota hasta. Et impellito / et infligito celeriter quacumque parte corporis ipsum laedere poteris.

Cod. Vinob. 10825, 164r

Ain unders und obers anbinden zue halber stangen.

Item halt dich also in disem yufechten bin ob er also von under und ober ain ander so trit / mit dein linggen fuoß hinein und stoß im mit dem hindern orth nach seinem gesicht sezt er / dir daß also ab so schlag im mit dem vordern ort nach den kopft.

Schlagt er dir also zue sez im / daß ab mit halb und gannzen stanngen und stoß und schlag behennd zue im wa dü in . . . / magst.

An upper and lower bind using half-staffs.

If you bind him from from above (as depicted by the left figure), and he binds you from below (as depicted by the right figure), then advance with your left foot and thrust your back point into his eyes. If he sets you aside, then strike his head with your front point.

If he strikes your head thusly, then set him aside with the half or whole staff, and quickly thrust or strike to whatever part of his body you can.

SHORTSTAFF 22

Cod. Vinob. 10825, 164v

Aversio cum percussione.

 Praefigas laevum pedem tuum, et dextro consequere, et utere impulsu, versus hostis visum, tum excipis hoc modo / ictum eius declinando.

 Si vero ille eodem modo excipiendi contra te utitur, tum progreditor laevo pede tuo et impellito / inferiorem mucronem versus visum eius. Si exceperit hoc tibi, prosite celeriter post ipsum, et / ferito caput ipsius et apprehendito celeriter hastam tuam denuo rursum, et impelle ictum longa hasta / versus eius visum.

Cod. Vinob. 10825, 164v

Ain absetzen gegen ainem schlag.

 Item schickn dich also in disem zufechten, stanndt mit deinem linggen fuoß vor und trit / mit deinem rechten hinnach und stoß im nach dem gesicht so nimbst du im sein stanng / ab.

 So er dir den streich also abwend so trit mit dem linggen fuoß hinnach and wind im / von unden auff nach seinem gesicht versezt er dir das so spring [...] inn das linggen / und schlag im nach dem habut und greiff behennd nach deinem stanngen widerumb und / stoß mit lanngen stanngen im nach dem gesicht.

A displacement against a strike.

Position yourself thusly during the approach: stand with your left foot leading, pass forward with your right foot, and thrust into the opponent's eyes while setting his strike aside.

If he sets your strike aside in this same way, then pass forward with your left foot and wind into his face from below. If he parries this, then quickly spring around him on the left side, strike his head, seize your staff again, and thrust into his eyes with the long point.

Chapter 3

Lance and Longstaff

"**LANCEA**, cuius nostra memoria maximus usus est, sed olim admodum / infrequens, continet habitus duodecim, una cum viginti quator formis" (Cod. Icon. 393, 14r, 13–14).

"The **LANCE**, our memory of the use of which is greatest, but in the future, is increasingly rare. [This chapter] contains 12 techniques, together with 24 illustrated pages."

The weapon in this chapter appears to be a staff roughly 10 feet in length. However, Mair identifies it by the Latin word *lancea*, or "lance." Though often associated with jousting, the term "lance" may denote any spear of considerably greater length than the typical infantry spear. In fact, Mair's German word for this weapon, *langenspiess*, translates to "long spear."

The lance saw widespread battlefield use in the ancient world. The *sarissa*, a lance upward of 13 feet in length, was incorporated into phalanx warfare with great success by Phillip II of Macedon and his son, Alexander the Great. Lances were later adopted as a cavalry weapon by the Romans and found continued use well into the Middle Ages. In the 15th century, the Swiss revived the use of the lance as an infantry weapon, dominating Europe with their impenetrable pike formations.

As in the preceding chapter, Mair's word choice and linear footwork suggest that this chapter deals with lance techniques. We conclude that the longstaff may be used as a training weapon for the spear, and that the two are largely interchangeable.

Like the spear, the lance is primarily used for thrusting, but it is much less suited as a striking weapon due to its length. Still, many of the techniques in the preceding chapter are applicable here.

Mair's terms for the parts of the longstaff or lance are identical to those for the shortstaff or spear.

LONGSTAFF 1

Cod. Vinob. 10825, 166r

Primi duo lancearum contactus de latere sinistro, cum loco eorum fortiori et infirmiore.

Hunc in modum te compones lancea athletice gubernanda, ex latere sinistro eam continebis, pedem sinistrum praepones, lanceae mucronem posteriorem manu dextra de latere dextro porrecti extendes, manum vero sinistram in latere / sinistro lanceae tuae adplicabis. Si itaque te adaptaris, intro procedes, pedemque dextrum prefigito, et faciem adversaris punctione adpetito.

Sin autem is idem contra te fuerit molitus sinistrum pedem praeponentem, atque iuxta genu sinistrum, / lanceam tuam manu sinistra contineas, nec non manu dextra iuxta pedem dextrum, mucronem alterum, impulsum hostilem avertes parte anterioris lanceae. Interim vero circumvoluendo hastam rotabis, et consequutus dextro, pectus hostis / de latere dextro in latus ipsius dextrum adgreditor. Verum si se defendendo eum impulsum averterit, lanceae partem anteriorem terrae inclinato, pede dextro recedes, lancea tua ante faciem sit directa, manumque dextram supra caput tuum ex ea levabis, / subito autem dextram tenentem lanceam lateri dextro adplicabis, veru dextro si rursus intro contra adversarium processeris de latere dextro versus adversaris sinistrum contra eius faciem lanceam impellito.

Caeterum si hostis eo modo te adgressus est, / inflectendo id avertes, atque dextrum pedem reducas, sursum lanceam si vibraris, concede in locum firmum, idest rursus lanceam in visum hostis porrigas, atque animadvertas, num fortiter vel minus resistat, in latus dextrum, et lanceam / rursus arripito in parte anterioris, dextro itaque pede insequutus, eam in faciem adversaris ex eius latere dextro propellito. Porro id si hostis averterit, dextro retrorsum concedas, interimque lanceae partem anteriorem, terrae inclinabis, atque / sursum ea sublata supra caput tuum in tu[is] defensione tuta, ab adversario recedes.

C93, 194r

Die ersten zway anpinden vonn der linncknen seyten mit der schwech und sterckn.

Item schickn dich allso mit disem zufechten von deiner lincknen seiten mit deinem lanngen Speiß seß deinen Lincknen fuoß fur und halt deinen / hindern orth in deiner rechten hand wol gestrackt hinauß aüf deiner rechten seiten dein lincknen hand aüf deiner lincknen seitten In deinem spieß Indes trit mit deinem rechten schenckel hinein und stoß Im / nach seinem gesicht.

Stost er dir dann also nach deinem gesicht unnd dü mit deinem lincknen fuoß vor steest dein lincknen hand bey deinem lincken knie In deinem spieß dein rechte hand an deinem rechten / schennckel deinen ort daran gescht seß er Im den stoß vornen an deinem spieß ab Indes wechsel Im durch unnd trit mit deinem recten fuoß hinach und stoß Im zü seiner prüst von deiner rechten aüf sein / rechten seiten seßt er dir das ab so seß deinen speiß vornen aüf die erden mit deinem Ort sincknen Indes trit mit deinem rechten schenckel zü rückn dein spieß vor deinem gesicht mit deiner rechten hand ob / deinem haüpt Indes wend dein rechte hand Inn dein rechte seitten mit deinem spieß trit mit der rechten füoß wider hinein und stoß im von deiner rechten auff sein lincken seiten zü seinem gesicht.

Stost Er / dir allso zü so wind im das ab von deinem spieß Indes trit mit deinem rechten schenckel zü rückn und schwind deinen spieß aüf Im dem trit Inn die sterckn aüf dein rechten seiten und greiff mit deiner / lincken hand widerumb vornen In dein spieß volg mit deinem rechten füoß hinnach und stoß Im aüf sein rechten seiten seiner gesichts verseßt Er dir das so seß deinen rechten füoß wider zü Rückn Indes / laß deinen spiess vornen bey dem ort aüf die erden sincknen und wind dich aüf in güte versaßung mit deinem spieß uber dein haüpt unnd zeüch dich damit zü Rückn.

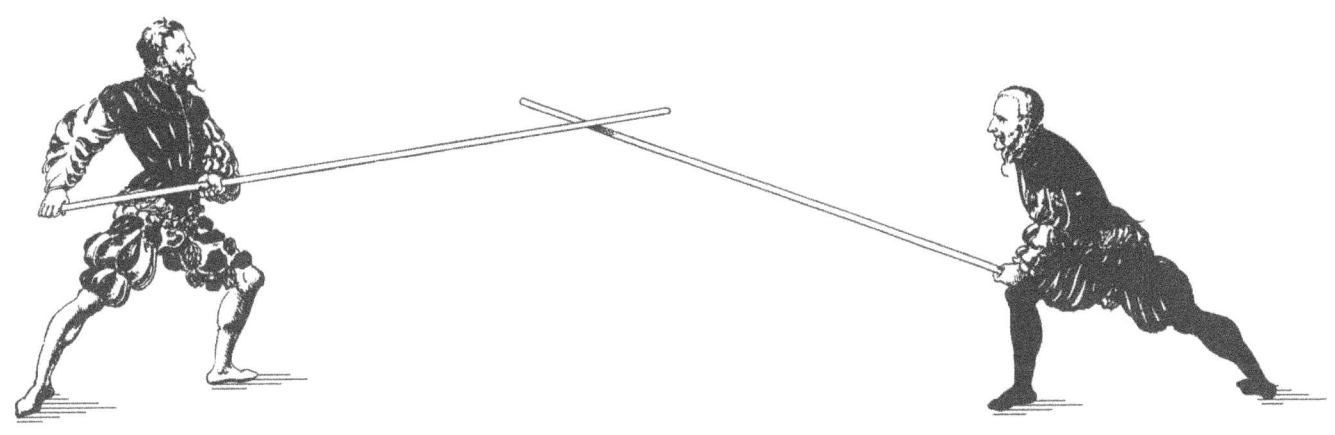

The first two binds from the left side with the weak and strong.

During the approach, position yourself as follows: hold the weapon on your left side, place your left foot forward, hold the back point in your right hand, extend the right hand out from your right side, and grip the longstaff with your left hand on the left side (as depicted by the left figure); then, immediately step to the inside with your right foot and thrust into the adversary's face.

If he thrusts at your face while you are leading with the left foot, holding your longstaff with the left hand near your left knee and the right hand near your right foot, at the back point, then set his thrust aside with your front point (as depicted by the right figure), change him through, step forward with your right foot, and thrust into his chest from your right side. If he sets this thrust aside, then turn your front point toward the ground, bring the right foot back, and with the longstaff in front of your face and your right hand above your head, quickly bring the right hand to your right side, step forward with the right foot, and thrust from your right side toward his left, into his face.

If he thrusts toward your face in this same way, then parry him by winding, bring the right foot back, raise your longstaff up high, transition into the strong on your right side, thrust toward his eyes, determine whether he is strong or weak in the bind, place your left hand in front again, pass forward with the right foot, and thrust from the right side into his face. If he parries you, then step backward with your right foot, simultaneously turn your front point toward the ground, and withdraw from the adversary with your weapon raised above your head in a strong guard.

LONGSTAFF 2

Cod. Vinob. 10825, 166v

Lanceae locus firmus et infirmus ex primo contactu.

Tum ad hostem lancea rite gubernando perveneris, dextro pede ad eum intro concedas, versumque faciem eius lancea directa sit, manus sinistra supra pedem sinistrum constituatur, eoque habitu pede sinistro si ad hostem intro processeris, collum eius / lanceae impetu adpetes.

Verum si adversarius te ex eodem habitu fuerit adgressus dextrum pedem praefigentem, manus tua dextra genu dextro adplicata, et manus sinistra lanceam contineris supra pedem sinistrum collocata, te itaque composito secundum / descriptionem modo factum, lanceae parte anteriori eius impulsum eludito de latere dextro tuo, atque subito dextro pede intro si processeris, visum eius adgreditor. Sin autem hostis tuam impulsum averterit, circumrotando hastam inferne pectus / impulsu adpetito. Interim vero dextro pede consequutus, lanceam elevando in statum liborum, status is est, si lanceam tenueris iuxta latus vel dextrum vel sinistrum capitis coa[da]ptabis, et visum adversariis lanceae impetu adgreditor.

Et / si ex eodem habitu hostis te adgreditur, sinistrum referto, et excipito ipsum mutatorio cancellato versus latus tuum sinistrum, curabisque ut etiam dextrum reducas, et sursum ex mutatorio lanceam inflectes versus latus dextrum, rursumque dextrum / introponito pedem, atque fortiter eius visum affliges. Sed si hostis tuum conatum fecerit irritum, lanceam si retraxeris rursus locum hostis visus praescriptum adpetes. Interea autem, locum lanceae anteriorem deorsum demittes, dextrum vero brachium / sursum supra caput tolles, et lanceam ante faciem tuam ex tuta tuis defensione continebis.

Caeterum, si adversarius contra te gemino impulsu lanceae fuerit usus, tum eius impetum avertere haud dubites intra manum tuam utramque, insequutus intro / sinistro pede, pectusque eius fueris adgressus lanceae impulsu, tuto ab adversario recedere curabis.

C93, 194v

Die schwech unnd sterch auß dem Anpund.

Item wann du mit dem zufecten zu dem Man kümpst so trit mit deinem rechten füoß gögen Im hinein halt deinen spieß gögen seinem / gesicht, mit deiner lincken hand ob deinem lincten schenckel Indes trit mit deinem lincken füoß hinein unnd stoß Im nach seinem hals.

Stost er dir also zü unnd dü mit deinem rechten fuoß vorstast dein rechte / hand ob deinem rechten knie dein lincken hinden ob deinem lincken schenckel Inn deinem spieß so seß Im der stoß ab vornen Inn deinem spieß aüf dein rechten seiten Indes trit mit deinem lincken schenckel hinein und / stoß Im nach seinem gesicht verseßt er dir das so wechsel Im unnden dürch unnd stoß im nach seiner prüst Indes volg mit deinem rechten füoß hinnach unnd gee mit deinem spieß aüf in den freystand und stoß Im / nach seinem gesict.

Stost er dir also aüß dem freystand zü so zeüch deinem lincken schenckel zü rückn unnd seß im das ab mit geschencken wechsel aüf dein lincken seiten Indes trit mit deinem rechten füoß aüch zü / rückn unnd wind dich aüf aüß dem wechsel aüf dein rechten seiten mit deinem spieß In dem trit mit deinem rechten füoß wider hinein und stoß Im mit der sterckn zü seinem gesicht verseßt er dir das so zückn deinem / spieß wider zü dir unnd stoß im behend widerümb zü der vorgeschribnen stat seiner gesicht Indes laß deinem spieß vornen nider mit deinem rechten Arm wol aüf uber dein haüpt deinen spieß vor deinein angesicht / Inn güter versaßung.

Stost er dir also zwifach zü deinem gesicht so nimb im dem stoß hinweckn zwischen deinen baiden henden Inn deinem spieß Indes volg mit deinem lincken schenckel hinach und stoß / Im nach seiner prüst wend dich damit zü rückn Inn güter versaßung.

Lance and Longstaff

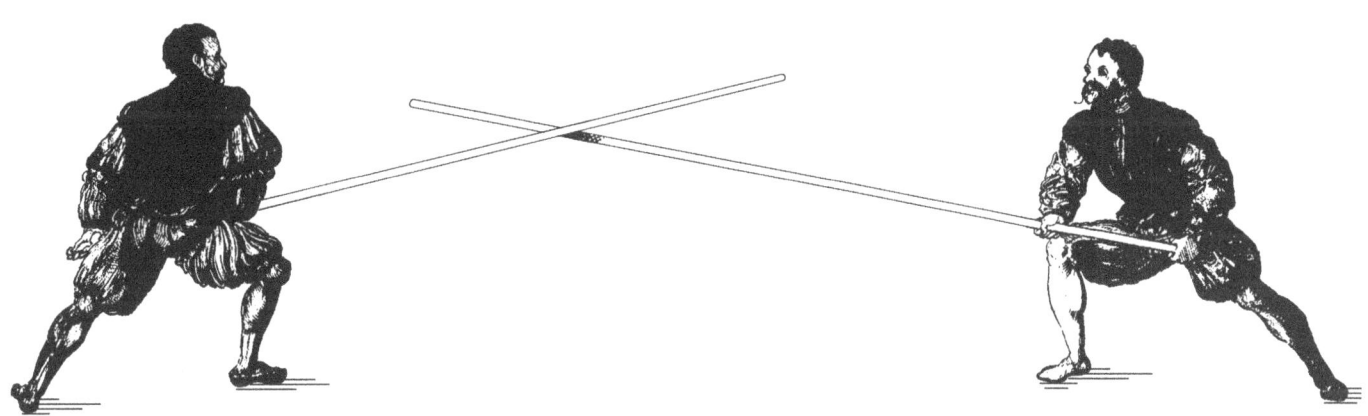

The strong and weak in the bind.

As you approach the enemy, step toward him with your right foot, aim your longstaff at his face, and position your left hand over your left leg (as depicted by the left figure); then, step to the inside with your left foot and thrust into his throat.

If he thrusts toward you while you are standing with your right foot leading, your right hand over your right knee, and your left hand above your left foot, then parry his thrust from your right side with the front of your staff (as depicted by the right figure), immediately step to the inside with your left foot,[1] and thrust into his eyes. If he parries your thrust, then change under him and thrust into his chest. At the same time, pass forward with your right foot into a free stance, raise your lance up, and thrust into his eyes.

If he attacks you thusly from a free stance, then bring your left foot back, set him aside with a crossed *wechsel*[2] on your left side, bring the right foot back, wind your longstaff from the wechsel onto your right side, step to the inside again with your right foot, and violently thrust into his face. If he parries your thrust, then bring your longstaff back, quickly thrust at his face again, lower your front point, raise your right arm high above your head, and hold the lance in front of your face in a strong guard.

If he attempts this same double thrust to your face, then parry him between your hands, simultaneously pass forward with your left leg, thrust into his chest, and withdraw from him in a strong guard.

---

1. The Latin text reads dextro pede, "left foot."
2. German, "changer." See Latin mutatorius, "change."

Longstaff 3

Cod. Vinob. 10825, 167r

Dextra et sinistra custodia aperta.

In custodiam dextram apertam hoc modo te accomodabis, dextrum pedem praepones, pars lanceae anterior terrae sit proclinata, manus sinistra iuxta mucronem posteriorem collocata, dextra vero iuxta genum dextrum, simulque intro procedes / pede sinistro, mucronem anteriorem tolles sursum, atque visum adversariis appetito.

Si vero is eadem contra te molitus fuerit praeponentem pedem sinistrum, manusque tua sinistra lanceae iuxta genu sinistrum adplicata, dextra vero / iuxta mucronem posteriorem posita, tum adversariis impulsum ex custodea aperta in custodiam cancellatam avertes, atque pedem sinistrum retrorsum reducas. Interea autem rursus inflectes sursum in habitum liberi status tibiis paribus, ita ut / lancea cubito tuo sinistro recumbat, insequutus subito pede sinistro, visum hostis animose impulsu adgreditor. Et si eum conatum averteris hostis, dextro insequitor, pectus ex latere eius sinistro adpetito.

Verum si eadem ratione te is adgreditur, tum ipsius impulsum / mucrone lanceae tuae anteriori, de tuo latere sinistro versum latus tuum dextrum, subito autem pede dextro consequutus, pudenda adversariis impulsu adpetes. Hoste conatum tuum irritum facere conante, sinistro consequitor, sursum supra caput lancea sublata, / eius faciem impulsu oblidito.

Quum vero is duplici impulsu fuerit usus, eum mucrone anterioris excipies, confestim autem circumvoluendo lanceam de adversariis latere sinistro, versus eius dextrum, impellito in eius visum vel pectus, atque dextro mox insequutus / fortiter et longe lanceam versus adversarium propellere poteris, et ex eo habitu te in custodiam apertam converte in latus tuum dextrum ita, ut pes dexter in ipsa conversione retrorsum vertatur, atque ea ratione rursum recedito ab adversario in tuta / tamen tuis defensione.

C93, 195r

Ainn rechte unnd ain linnckne offne hut.

Item schickn dich also mit disem zufecten Im die recht offenhut stannd mit deinem Rechten füoß vor mit deinem spieß vornen / aüf der erden dein lincken hannd bey deinem hindern orth dein rechte bey deinem rechte knie Indes trit mit deinem lincken schenckel hinein gee vornen mit deinem spieß ubersich unnd stoß Im zü seinem gesicht.

Stost er dir dann also dem gesicht zü unnd dü mit deinem lincken füoß versteest dein lincke hand vornen bey deinem lincken knie in deinem spieß dein rechte hinden bey deinem Ort so seß im seinen stoß ab / aüß der offnenhut Inn die schrannckhut unnd seß deinen lincken füoß zü rückn Indes wind dich widerumb aüf Inn den freystannd mit gleichen füoßen zü sainen daß dein spieß aüf deinem lincken elnpogen / sig Inn dem volg mit deinem lincken schenckel hinnach unnd stoß Im nach seinem gesicht verseßt Er dir daß so volg mit deinem rechten füoß hinnach unnd stoß Im nach seiner lincken prüst.

Stost er / dir also zü so nimb im seinen stoß ab mit deinem lanngen Ort von deiner lincken aüf dein rechten seiten Indes volg mit deinem rechten schenckel hinnach unnd stoß Im nach seinen gemechten verseßt Er / dir das so volg mit deinem lincken füoß hinnach gee mit deinem spieß wol aüf uber dein haüpt unnd stich im nach seinem gesicht.

Stost er dir also zwifach zß so seß Im das ab mit deinem vorderen orth deiner / spieß Indes wechsel Im durch von seiner lincken seiten auf sein rechte zu seinem gesicht oder der prüst unnd volg mit deinem rechten schenckel hinnach so hastü ainn lanngen stoß In dem wind dich Inn die offenhut / aüf dein rechte seitten das dein rechter füoß mit dem winden zü rückn gee unnd wend dich damit widerumb zü rückn Inn güte versaßung.

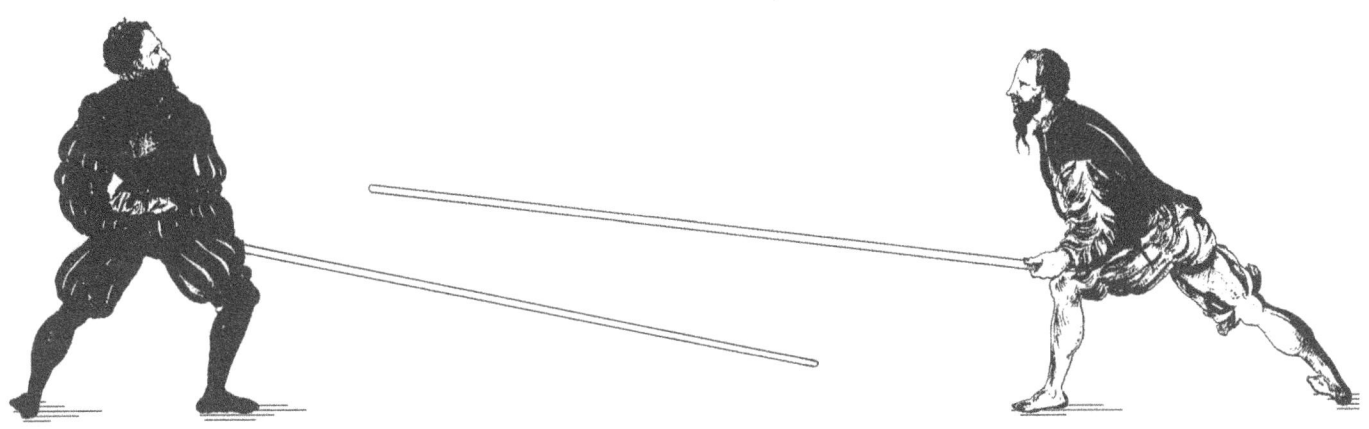

The right and left open guard.

Position yourself thusly for the right open guard: stand with your right foot leading, tilt your front point toward the ground, place your left hand near the back point, and place your right hand near the right knee (as depicted by the left figure); then, advance to the inside with your left foot, raise your front point, and thrust into the adversary's eyes.

If he thrusts at your face while you are standing with your left foot leading, your left hand by your left knee, and your right hand holding the longstaff at the back point (as depicted by the right figure), then set his thrust aside from open guard into *schrankhut*,[1] and bring your left foot back. At that moment, wind upward into a free stance with both legs equally toward him, rest the longstaff on your left elbow, immediately pass forward with your left foot, and thrust into his eyes. If he parries you, then pass forward with your right foot and thrust from his left side into his chest.

If he thrusts at you in the same way, then set him aside from left to right with your long point, immediately pass forward with your right foot, and thrust into his groin. If he parries you, then pass forward with your left foot, raise the longstaff above your head, and thrust into his face.

If he thrusts twice at you thusly, then set him aside with your front point, immediately change him through from his left side to his right, thrust into his face or chest, pass forward with your right foot, extend a long thrust, bring your right foot back as you transition into an open guard on your right side, and withdraw from him in a strong guard.

---

1. German, "barrier guard." See Latin *custodia cancellata*, "crossed guard."

LONGSTAFF 4

Cod. Vinob. 10825, 167v

Status liberi habitus contra custodiam cancellatum.

Quum ad hostem lancea more athletico quassata porveneris, hunc in modum te in habitum liberi status accomodabis, erectus corporis habitu consistas ita, ut latus sinistrum sit porrectus, supra brachium sinistrum lancea contra adversarium / promineat, manus vero dextra mucronis posterioris sit applicata, insequutus confestim pede dextro, pro viribus lanceam corpori hostis iniscito.

Verum si is eodem modo te adgressus fuerit in mutatorio cancellato consistentem atque contra / adversarium pedem dextrum praeponentem, dextrum reducere memineris retrorsum, et impulsum ipsius eludes, aversa itaque hostis lancea, deflectes in latus tuum dextrum. Verum in ipsa aversione si dextro pede fueris consequutus, visum adversaris impulsu / sauciato. Sin autem is tuum impulsum averterit, sinistro consequitor, et circumvoluendo de latere tuo dextro versus hostis sinistrum hastam pectoris eius infigito.

Caeterum si adversarius idem contra te moliatur, mucrone lanceae anteriori / excipere eius impetum non dubites, interim vero insequutus pede sinistro, pectus hostis pungendo quassabis. Sed si is eum impulsum exceperit, tum de latere dextro versus eius sinistrum lancea circumvoluta inferne locum eundem adpetes.

Verum / si is dupliciter eadem ratione te fuerit adgressus, sinistrum pedem praeponentem, eum impetum parte priori lanceae avertes, statim vero dextro consequitor pede, et si visum eius adgressus sis, dextrum pedem reduces, terrae lanceam proclinate / de parte anterioris, atque supra caput tuum manu dextra sublata, lancea ante faciem tuam consistat curabis. Verum si contigerit ut adversarius instando se sequatur, tum longe, quantum poteris lanceam contra eum propellito, et retrorsum / deflectes in tuis tuta defensione.

C93, 195v

Der freystannd gögen der schrannkhut.

Item wann du mit dem zufechten zu Dem mann kumpst so halte dich also mit dem Freistannd stee aüf recht mit deinnem / leib dein lincke seiten fürgewenndt deinen spieü gögen dem man aüf deinem lincten arm dein reckte hannd gestract hinden bey deinem Ort Indes volg mit deinem rechten schennckel / hinnach unnd stoß im gewaltig zü seinem leib.

Stost er dir also zü unnd dü Inn dem geschrenctin wechsel steest deinem recten fuosß furgeseßt gügen dem man so trit mit deinem rechten füöß zü / rückn unnd seß Im seinen stoß ab wend dich damit aüff dein rechte seiten unnd In dem abseßen so volg mitt deinem rechten füöß hinnach unnd stich Im nach seinem gesicht verseßt er dir das so trit / mit deinem lincken schenckel hinnach unnd wechsel Im durch von deiner rechten aüf sein lincken seiten stoß Im du mit zü seiner prüst.

Stost Er dir also oben zü so seß Im das ab mit deinem vordem / ort deiner spieß Indes volg mit deinem lincken schenckel hinnach unnd stoß Im nach seiner prüst seßt er dir den stoß ab so gee Im unnden dürch von seiner rechten aüf sein lincke seiten unnd stich Im / damit zü derselbigen stat.

Stost Er dir also zwifach zü unnd dü mit deinem lincken füöß vorsteest so seß Im den stoß ab vornen an deinem spieß Indes volg mit deinem rechten schenckel hinnach und / stoß Im zü seinem gesicht unnd zeüch deinen rechten schenckel wider zü rüctn deinen spieß vornen aüf der Erd unnd fur mit deiner rechten hand wol aüf ober dein haüpt das dir der spieß vor deinem / gesicht stee raist er dir dann nach so stoß ainem langen stoß aüf In mit deinem spieß wend dich da mit zü rückn in güte versaßung.

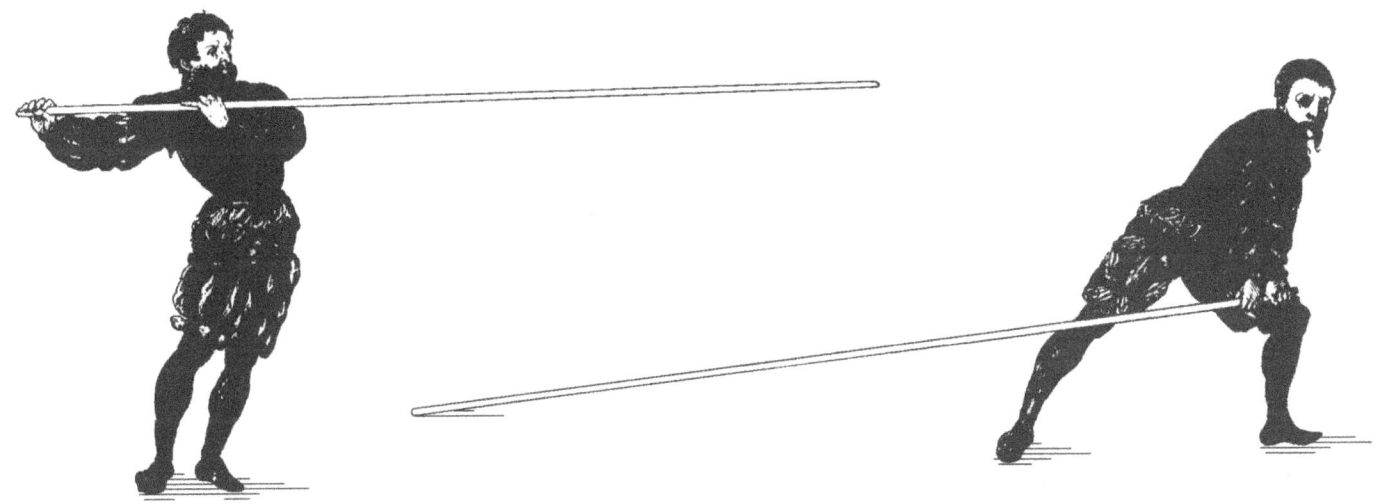

Longstaff 4

The free stance against *schrankhut*.[1]

During the approach, position yourself in a free stance thusly: stand in an upright position with your left side leading, your longstaff extended toward the adversary over your left arm, and your right hand extended by the back point (as depicted by the left figure), then immediately pass forward with your right foot and thrust into his body with all your might.

If he thrusts thusly toward you while you are standing in the crossed guard with your right foot leading (as depicted by the right figure), then bring your right foot back, set his thrust aside, wind onto your right side, pass forward with your right foot, and thrust into his eyes. If he parries your thrust, then follow with your left foot, change him through from your right side onto his left, and thrust into his chest.

If he thrusts at you in the same way, then set him aside with your forward point, simultaneously pass forward with your left foot, and thrust into his chest. If he parries your thrust, then change him through from his right to his left side and thrust to the same target.

If he thrusts twice toward you while your left foot is leading, then set his thrust aside with your front point, simultaneously pass forward with your right foot, thrust into his eyes, bring your right foot back, tilt your front point toward the ground, and hold the longstaff in front of your face with your right hand raised above your head. If he pursues you, then thrust your long point as far as you can toward him, and step back into your strong guard.

---

1. German, "barrier guard." See Latin *custodia cancellata*, "crossed guard."

LONGSTAFF 5

Cod. Vinob. 10825, 168r

Contactus hastarum de medio, contra impulsum, quo visus hostis adpetitur.

In eum habitum hoc modo te compones, sinistrum pedem praepones, manum utramque lanceae medietati adplicabis, sinistra tamen anterior sit, dextra iuxta sinistrum pedem posita, atque ex ea forma per manus lanceam versum hostis pectus impellito.

Verum si eodem habitu is fuerit usus contra te, in impulsu, quo visus leditur, consistentem, sinistrumque pedem praefigentem, dextro consequitor, lanceam versus latus tuum sinistrum inflectito, atque ex ea forma visum adversariis figito. Sin autem / is avertere conatus est, pedem dextrum referto. Sed si tibi adversarius insequendo institerit contra te lanceam propellendo, tum te in cancellatum mutatorium in latus tuum dextrum deflectes, et impulsum ipsius eludes. Verum rursus dextro / contra eum intro si processeris, pectus seu visum hostis adgredere impulsu lanceae.

Cum vero is gemino impulsu vicissim te adpetat, parte anterioris lanceae eum repelles, interimque circumvoluendo lanceam de latere tuo dextro versum adversariis / sinistrum, axillam ipsius sinistram impulsu quassabis. Excipiente id hoste, sinistro pede insequutus pudenda eius figito.

Cum vero ipse eodem modo fuerit contra te usus dextrum praeponentem pedem, eum retrahere memineris, atque / impulsum hostis vitabis avertendo in latus tuum sinistrum, confestim vero rursus introgreditor, et praefixum hostis pedem lanceae mucrone adpetito. Volente autem eo impetum hunc eludere, hasta celeriter sublata, et directu versus ipsius / visum, in tuta tuis corporis defensione retrorsum concedes.

C93, 196r

Ein anpinden mit dem halben spieß gögen ainem gesict stoß.

Item schickn dich allso mit Disem zufechten Inn den anpundt mit dein halben spieß stee Mit deinem lincken füoß vor dein spieß Inn / der mite mit baiden henden das dein lincke hand vornen gewendt sey die recht bey deinem lincken schenckel Indes scheüß im deinen spieß durch deine hend zü seiner prüst.

Stost er dir also zü unnd dü In / dem gesicht stoß gögen Im steest mit deinem lincken föoß vor so volg mit deinem rechten schenckel hinnach unnd wind mit deinem speiß aüf dein lincke seiten unnd stoß im damit zü seinem gesicht / verseßt Er dir das so zeüch deinen rechten schenckel wider zü rückn raist er dir dann nach mit ainem stoß so wend dich In den geschrenckten wechsel aüf dein rechten seiten und nimb Im den stoß hinweckn / Indes tris mit deinem rechten schenckel wider hinnein unnd stoß Im nach seinem gesicht oder der prüst.

Stost Er dir also zwifach nach deinem gesicht so seß Im das ab vornen an deinem spieß Indes wechsel / Im dürch von deiner rechten seiten aüf sein lincke unnd stoß im nach seiner lincke Achsel verseßt er dir das so volg mit deinem lincken schenckel hinnach unnd stich ~~unnd stich~~ Im nach seinen gemechten.

Stost er dir also unden zü deinen gemechen unnd dü mit deinem rechten füoß vorstest so seß den zü rückn unnd wend Im den stoß ab aüf dein lincke seiten Indes trit behend widerümb hnnein unnd stoß Im / nach seinem fur geseßten schenckel […] er dir den stoß abseßen so gee behend wider aüf Im zü dem gesicht unnd zeüch dich damit zü rückn inn güter versaßung.

A bind with the half-staff against a thrust to the face.

For this technique, set your left foot forward, hold your longstaff in the middle with both hands so that your left is leading and your right is near the left leg (as depicted by the left figure), and thrust through your hands into the enemy's chest.

If he thrusts toward you in this way while you are standing upright in a face-thrust with your left foot leading (as depicted by the right figure), then pass forward with your right foot, wind the longstaff onto your left side, and thrust into his eyes. If he parries you, then bring your right foot back, and when he advances and thrusts toward you, transition into *wechsel*[1] on your right side and parry his thrust. At the same time, step toward him again with your right foot and thrust into his eyes or chest.

If he thrusts twice toward your face, then set him aside the forward part of your longstaff, simultaneously change him through by winding from your right side toward his left, and thrust into his left flank. If he parries your thrust, then pass forward with your left foot and thrust into his groin.

If he thrusts thusly toward your groin while your right foot is leading, then bring your right foot back, set his thrust to your left side, immediately step forward again, and thrust into his leading leg. If you want to set his thrust aside, then quickly lift your longstaff up, point it toward his face, and withdraw from him in a strong guard.

---

1. German, "changer." See Latin *mutatorius*, "change."

LONGSTAFF 6

Cod. Vinob. 10825, 168v

Impulsus per quem visus adpetitur, contra eum, quo pudenda adpetuntur.

Quum lancea athletice regenda ad hostem perveneris, intro concedes pede sinistro, atque de latere tuo sinistro versum latus eius dextrum visum adversariis lanceae mucrone oblidito.

Verum si is te eodem modo adgreditur contra eum pede / sinistro consistentem, mucrone anteriori eum profliges. Verum interea eius pudenda adgredieris.

Si hostis is idem contra te usurparit, pedem sinistrum referto, et impetum eius parte anterioriis lanceae repelles, rursus autem sinistrum intro si direxeris / pedem, pectus adgreditor lanceae mucrone. Ceterum si is defendendo se impulsum repulerit, recta ab eius lancea inflectes sursum tuam contra hostis visum.

Sin vero ipse usus fuerit impulsu gemino, tum lanceae medio avertere non / dubites. Interea autem pede dextro consequutus, vicissim eius visum lanceae, mucrone quassabis. Sed si eum impulsum animadverterit, eluseritque, lancea circumvoluta, pede sinistro consequitor, et lanceae mucrone latus adversariis / dextrum tundito.

Caeterum si is idem contra te fuerit molitus praeponentem pedem sinistrum, tum parte lanceae anterioris ex latere tuo dextro avertes hostis impetum, in ipsa vero aversione pede dextro intro procedes, atque de latere adversariis / dextro pectus fodito. Excipiente id hoste dextrum reducas pedem, et te in habitum fortudinis componas lancea, idest contra faciem adversariis lanceam porrigas, sentiasque num fortiter eam teneat. Ex ea itaque forma cum adversario certabis / aliam ex alia nuditate quaerendo, tum superne tum interne, ubicumque nudus repertus fuerit.

Sin autem tuus nuditates ratione eadem conquisierit hostis, sublata lancea supra caput tuum, eius impetum avertes, interea autem visum eius adgreditor, / atque ab eo recedito, id autem ut tuto fiat curabis.

C93, 196v

Ain gesicht stoß gögen ainem gemecht stoß.

Item wann du mit dem zufechten Yu dem man kumpst, so trit mit deinem lincken schenckel hinein und stoß Im von deiner lincken seiten auf sein / rechte zu seinem gesicht.

Stost er dir also nach deinem gesicht unnd dü mit deinem lincken fuoß vorsteest gügen Im so seß Im seinen stoß ab mit deinem vordern orth Indes stoß im nach seinen gemechten.

Stost er dir / also nach deinen gemechten so seß deinen lincken schenckel zü rückn unnd seß Im den stoß ab vornen an deinem spieß Indes trit mit deinem lincken schenckel wider hinnein unnd stich Im nach seinem leib verseßt / Er dir das so wind im gestrach an seinem spieß auf zu seinen gesict.

Stost Er dir dann also zwifach zu so verseß Im das mitten In deinem spieß Indes volg mit deinem rechten schenckel hinnach unnd stoß / Im auch nach seinem gesicht wirt er des stoß gewar unnd seßt dir den ab So wechsel durch volg mit deinem lincken fuoß hinnach unnd stoß Im nach seiner rechten seiten.

Stost er dir also nach deiner rechten seiten / unnd dü mit deinem lincken fuoß vorsteest so seß Im den stoß ab vornen an deinem spieß aüf dein recht seiten unnd in dem ab seßen trit mit deinem rechten sch[enkel] hinnein unnd stoß Im nach seiner prüst / aßf sein recht seiter verseßt Er dir den stoß so trit mit deinem rechten füoß wider zü rückn unnd gib dich mit deinem spieß Inn die sterckn Indes arbait mit im vor einer ploß zü der anndern unden und / oben wa er dir werden mag.

Sucht er dir deine plossen also so gee mit deinem spieß wol aüf uber dein haüpt unnd seß im das ab Indes stoß Im nach seinem gesicht und zeüch dich damit zü rückn in gute / versaßung.

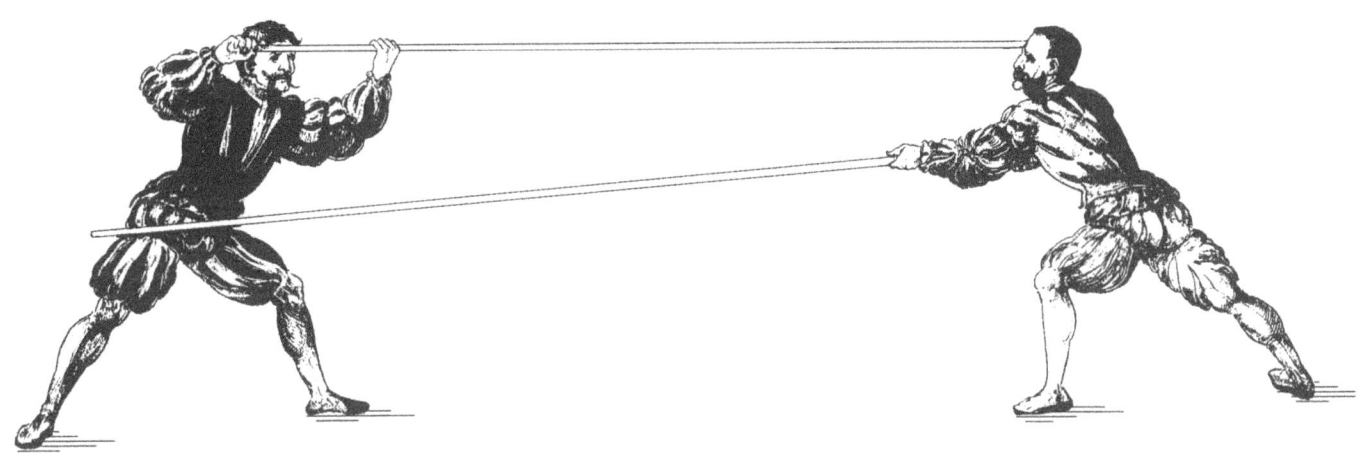

A thrust to the face against a thrust to the groin.

As you approach the enemy, step to the inside with your left foot and thrust from your left side toward his right, into his face (as depicted by the left figure).

If he thrusts to your face in this same way while you are standing with your left foot forward, then set his thrust aside with your front point and simultaneously thrust into his groin (as depicted by the right figure).

If he thrusts thusly at your groin, then bring your left foot back, set his thrust aside with the forward part of your longstaff, immediately step to the inside again with your left foot, and thrust into his chest. If he parries your thrust, then wind straight from his staff into his eyes.

If he thrusts twice toward you, then set him aside with the middle of your staff, simultaneously pass forward with your right foot, and thrust into his eyes. If he anticipates your thrust and sets you aside, then change through, pass forward with your left foot, and thrust into his right flank.

If he thrusts thusly to your right flank while your left foot is leading, then parry his thrust with the forward part of the staff from your right side; meanwhile, step to the inside with your right foot and thrust from his right side into his chest. If he parries your thrust, then bring your right foot back, position yourself with the longstaff in a strong stance—that is, extend your staff toward his face and feel if he is holding his weapon strongly—and probe him from one opening to the other, high and low, wherever his openings may be.

If he seeks out your openings in this same way, then raise your longstaff above your head, set him aside, simultaneously thrust into his eyes, and withdraw from him in a strong guard.

Longstaff 7

Cod. Vinob. 10825, 169r

Habitus mutatoriis simplicis contra custodiam de media lancea.

Sic te in simplicem mutatorium componas, sinistrum pedem praefiges, mucro lanceae terrae sit inclinatus, manus sinistra iuxta genu sinistrum lanceam continens, dextra vero iuxta tuum latus dextrum prope mucronem posteriorem constituatur / atque eo habitu pede dextro consequutus, latus adversariis sinistrum figito.

Sed si idem contra te moliatur hostis, consistentem in custodia mediae lanceae, sinistrumque pedem praefigentem, et manu utraque prope pectus lanceam continentem de latere / dextro, pede dextro insequutus, lanceae medio hostis conatum falles avertendo. Interea autem ad eum intro procedas, atque lanceam mediam versus eius visum impellito, hoc itaque si usus fueris, omnis eius conatus irritus fiet.

Verum si instando / atque urgendo te ratione eadem persequitur, et tu mucrone anterioris nihil ex hostis instantia poteris perficere, tunc per manum sinistram, lanceam retrorsum voluendo si torseris, rursum laborare et cum adversario concertare quibis, atque vicissim / ad eum prosilies, et lancea media habitu hostis adgreditor nuditatem proximam, ubicumque nudus adparverit. Sed si id averterit, sinistro pede insequutus, pectus adversariis punges.

Verum si is eadem habitu tuas nuditates conquisiverit, tum / dupliciter retrorsum recedes in habitum cancellatis mutatoriis, lanceam in manum dextram recipias iuxta mucronem posteriorem quam longissime poteris, atque cum hoste certabis ex loco firmioris et informioris aliam nuditatem ex alia conquirendo. Porro si / tuos impulsus eluserit repellendo, de manu dextra lanceam longissime versus visum hostis torquebis, interea autem retrorsum concedas quam tutissime poteris.

C93, 167r

Der ainfach wechsel gögen ainner hut Inn dem halben spieß.

Item schickn dich allso mit disem Zufechten in dem ainfachen wechsel stee mit deinem Lincken fuoß vor dein spieß auf der Erden / mit deinem Ort dein lincke hand bey deinem lincken knie Inn deinem spieß dein rechte bey deiner Rechte seiten hinden bey deinem Ort Indes volg mit deinem rechten schenckel hinnach / unnd stoß Im nach seiner lincken seiten.

Stost er dir also nach deiner lincken seiten unnd dü In der hut steest mit deinem halben spieß deinen lincken füoß fürgeseßt deinen spieß mit baiden henden / vornen an deiner prüst auf deiner rechten seiten so trit mit deinem rechten schenckel hinnach unnd seß im das ab mit deinem halben spieß Indes spring zü im hinnein und laß im deinnen / halben spieß zü seinem gesicht schiessen so mag er zü kainer arbaukumen.

Raist er dir also nach unnd dü mit deinem vordern Ort mit arbaiten kanst so laß den spieß zü rückn schiessen / durch dein lincke hand so magstu mit widerumb zu gleicher arbait kumen Inn dem spring auch widerumb zu im hinein unnd arbait mit deinem halben spieß zu seiner nechsten ploß / wa er dir werden mag seßt er dir das ab so volg mit deinem lincken füoß hinnach unnd stoß im nach seinem leib.

Sücht er dir deine plossen also mit seinem halben spieß so zeüch dich zwiffach zü / rückn inn den geschrenckten wechsel unnd nimb deinen spieü widerumb In dein rechte hand hinden bey deinem Ort auf das lengst Indes arbait mit Im schwech unnd sterckn von ainer ploß / zü der anndern wa Er dir werden mag verseßt er dir das so stoß Im mit deiner reichten hand nach seinem gesicht ainen lanngen stoß mit deinem spieß und zeüch dich damit zü rückn in / guotte versaßunng.

The simple *wechsel*[1] against a guard from the half-staff.

To position yourself in the simple *wechsel*, stand with your left foot forward, your front point tilted toward the ground, your left hand gripping the longstaff near your left knee, and your right hand on your right side, near the rear point (as depicted by the left figure); then, immediately pass forward with your right foot and thrust into the enemy's left flank.

If he thrusts thusly at your left flank while you are holding your longstaff in front of your chest in a half-staff guard on your right side and leading with your left foot (as depicted by the right figure), then pass forward with your right foot and set him aside with the middle of the staff. In that same instant, advance toward him on the inside and thrust your half-staff into his eyes; if you do this, all his efforts will be in vain.

If he closes in and presses toward you in this same way and you are unable to use your front point, then slide the longstaff back through your left hand so that you can work again, rush toward him, and wind your half-staff into his closest opening, wherever it may be. If he sets you aside, then pass forward with your left foot and thrust into his chest.

If he seeks out your openings thusly with his half-staff, then step back twice into a crossed *wechsel*, bring your right hand as close to the rear point as you possibly can, and fight him at the strong and weak, from one opening to the other. If he parries your thrusts, then extend a long thrust into his eyes with your right hand and withdraw from him in a strong defense.

---

1. German, "changer." See Latin *mutatorius*, "change."

LONGSTAFF 8

Cod. Vinob. 10825, 169v

Impulsus conversus contra habitum aversionis.

Hoc modo, ut in sequentibus manifestabitur, in impulsum conversum te compones more athletico. Pedem sinistrum praepones, manus sinistra lanceam continens iuxta genum sinistrum, dextra vero interpedem utrumque prope mucronem / posteriorem sit collocata ita, ut mucro anterior contra adversarium sit porrectus, te itaque consistente, dextro consequitor, et quasi eius latus sinistrum concutere velis, te accommodabis, verum conversis manibus, seu mutatis, / lanceam dextro hostis lateri infigito ex impulsu converso.

Caeterum si is idem adversum te fuerit molitus in aversione habitu consistentem, pedem sinistrum praefixeris, lanceamque de parte eius anterioris terrae inclinaris, et mucronem posteriorem / ante faciem sustuleris brachiis directis, tum superne et inferne de alio latere in aliud eius impetum avertes: eo igitur repulso, pede dextro consequitor, atque in latus tuum dextrum lancea sursum inflexa collum adversariis / impulsu quassabis. Sed si se ab eo impulsu defenderit, lancea circumvoluta de latere tuo dextro versum hostis sinistrum, visum ipsius adgreditor. Verum si is id animadverterit, averteritque, insequutus pede sinistro latus eius sinistrum / impulsu lanceae adpetito.

Sin vero eo habitu contra te sit usus dextro innitentem, tum manum dextram in latus sinistrum si inflexeris, hostis impulsus irritus fiet. Interea autem pedem dextrum reduceas, et rursum lanceam in latus tuum / sinistrum, in locum firmiorem accomodabis, atque rursus dextro insequutus, nuditatem supernam impulsu si tentaris, ab adversario in tuta tuis defensione regreditor.

C93, 197v

Ain verkerter stoß gügen ainem abnemen.

Item schickn dich allso mit disem Zufechten Inn den verkerten stoß stee mit deinem lincken fuoß vor Dein lincke hand vornen bey / deinem lincken knie Inn deinem spieß dein rechte zwischen deinen baiden schencklen hinden bey deinem Ort deinen vordern ort gügen seiner prüst gewendt Indes trit mit deinem rechten schencknel / hinein unnd ... wollestu Im zu seiner rechten seiten stossen Indes verker diene hend unnd stoß Im zü seiner lincken seiten mit ainem verkerten stoß.

Stost Er dann also gögen dir unnd dü In / dein abnemen steest deinen lincken füoß furgeseßt deinen spieß vornen auf der erden Inn der hoch vor deinem gesicht mit gestrackten Armen so nimb Im das ab unden unnd oben von ainer seiten zü der / andern Inn dem volg mit dem rechten schenckel hinnach unnd wind dich auf mit deinem spieß In dein rechte seiten unnd stoß Im nach seinem hals verseßt er dir das so wechsel durch von deiner / rechten auf sein linke seiten unnd stoß Im damit zü seinem gesicht wirt er des stoß gewar unnd seßt dir den ab so volg mit deinem lincken hinnach unnd stoß Im nach seiner lincken seiten.

Stost / Er dir also zü unnd dü mit deinem rechten füoß vorsteest so wind dein rechte hannd auf dein lincken seiten so seßstu Im den stoß damit ab Indes zeüch deinen rechten schenckel zü rückn und wend / dich mit deinem spieß widerumb auf dein rechte seiten In die sterckn Indes volg mit deinem rechten schenckel wider hinnach unnd stoß Im seiner obern ploß zü und zeüch dich damit zü rückn in / güte versaßung.

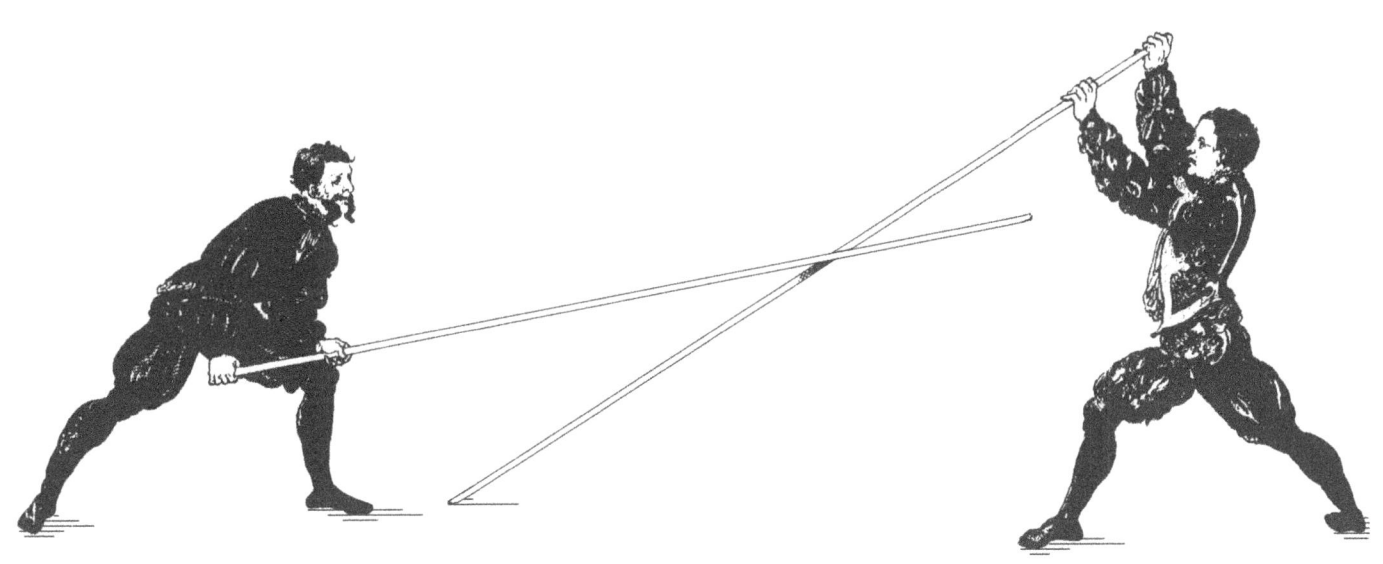

An inverted thrust against an *abnemen*.[1]

Position yourself thusly for an inverted thrust: place your left foot forward and hold your staff with the left hand near your left knee and your right hand between both of your legs, near the rear point, with the forward point extended toward the opponent (as depicted by the left figure), then immediately pass forward with your right foot, as if you intend to strike his left side, but instead invert your hands and thrust into his left flank.

If he attempts an inverted thrust against you when you are standing in a parry with your left foot forward, your front point lowered to the ground, and the rear point held in front of your face with your arms extended (as depicted by the right figure), then set him aside from above and below, from one side to another; next, pass forward with your right foot, wind the longstaff onto your right side, and thrust into his throat. If he sets your thrust aside, then change through from your right side to his left and thrust into his eyes. If he anticipates this thrust and parries, then pass forward with your left foot and thrust into his left flank.

If he thrusts toward you when you are leading with the right foot, then wind your right hand onto your left side, set his thrust aside, immediately bring your right foot back, bring the staff to your left side again at the strong, pass forward again with the right foot, thrust into his upper opening and withdraw from him in a strong guard.

---

1. German, "taking off." See Latin *habitus aversionis*, "parry; setting-aside."

LONGSTAFF 9

Cod. Vinob. 10825, 170r

Superna incursio, contra habitum supernae aversionis.

Quum ad hostem lancea rite gubernanda perveneris, pedem sinistrum praeponito, mediam lanceam arripe, mucro longior retrorsum propendeat, brachia egregie in lancea porrecta sint, manus sinistra anterior, dextra vero post caput / si collocata fuerit, mucrone anterioris intra brachium adversariis utrumque pectus pulsabis.

Sed si is eadem ratione obviam processerit tibi, pedem sinistrum praeponenti, atque pectus tuum adpetiverit, sinistrum reducito, et impulsum intra brachium / tuum utrumque eludito. Verum subito dextro etiam retrorsus resilias, et celeriter hastam per manus revoluas ita, ut mucro anterior in manu sinistra consistat, et eundem in visum vel pectus hostis infigito. Sin autem is averterit, et tibi institerit / persequendo, dupliciter circumvoluta lancea retrorsum concedas ita, ut eius mucronem posteriorem manu dextra contineas in latere dextro.

Porro si hostis eadem usurparit contra te, tum lanceam porrectum versus pudenda eius impellas, / atque dextro pede insequutus pectus adversariis fodito. Sed si impetum eum averterit, protendente te lanceam, eamque si circumvolueris de latere eius sinistro versus dextrum, infaciem hostis retorqueas. Avertente autem hoste eum impulsum, sinistro / insequutus, latus sinistrum eius adgreditor, simulque tuto ab hoste recedes.

C93, 198r

Ein obers ein lauffen gögen ainem Obern absetzen.

Item wann du mit dem Zufechten zu dem mann kumpst so schickn dich allso mit deine spieß seß deinem lincken fuoß fur unnd nimb deinen / spieß Inn der mitte laß deinen lanngen Ort zü rückn hanngen deine arm wol gestrach Inn deinen spieß die linck hand davorn die recht hinder deinem haupt Indes stoß Im mit deinem / vordern ort zwischen seinen baiden armen hinnein zü seiner prüst.

Ist er dir also ein gelauffen unnd dü mit deinem lincken füoß vorsteest unnd stoü dir zu deiner prüst so zückn deinen lincken / schenckel wider zü rückn unnd seß Im den stoü ab zwischen deinen baiden armen mit deinem spieß Indes spring behend mit deinem rechten füoß aüch zü rückn unnd laß deinen spieß behend / dürch deine hend schiessen das dü den vordern ort In deiner lincken hand habest Indes stoß im mit deinem vordern Ort zü seinem gesicht oder der prüst verseßt er dir das unnd raist dir nach / so wechsel dich behend zwifach zü rückn das dü deinen spieß widerumb bey deinem hindern ort habest in deiner rechten hand aüf deiner rechten seiten.

Wechselt er sich also von dir zü rückn / so scheuß im mit der lannge deiner spieß zü seinem gemechten Inn dem volg mit deinem rechten schenckel hinnach unnd stoß Im zü seiner prüst seßt er dir das ab unnd dü dein spieß wider / In die lennge gefast hast so wechsel Im durch von seiner lincken aüf sein rechten seiten zü seinem gesicht wirt er des stoß gewar unnd verseßt dir den so trit mit deinem lincken schenckel hinnach / unnd stoß Im nach seiner lincken seiten unnd zeüch dich dann zwifach zü rückn Inn güter versaßung.

A high charge-in against a high parry.

As you approach the opponent, place your left foot forward and grip your longstaff in the middle, with the long point angled backward, your arms well-extended on the lance, your left hand forward, and your right hand behind your head (as depicted by the left figure); then, immediately thrust your front point between his arms and into his chest.

If he charges toward you and thrusts at your chest in the same way while you are standing with your left foot forward, then bring your left foot back, set his thrust aside between your arms, simultaneously leap backward with your right foot, slide the spear through your hands so that the front point is in your left hand, and thrust it into his eyes or chest. If he parries and presses toward you, then pass backward while changing the longstaff twice, so that you are holding its back point with the right hand on your right side.

If he does the same thing to you, then thrust your long point into his groin, pass forward with your right foot, and thrust into his chest. If he sets your thrust aside, then while you are extended, change him through from his left side to his right and into his face. If he anticipates your thrust and parries, then pass forward with your left foot, thrust into his left flank, and withdraw from him in a strong guard.

LONGSTAFF 10

Cod. Vinob. 10825, 170v

Inferna incursio contra infernam aversionem.

In congressu contra hostem hac ratione te compones, si prope accesseris, partem lanceae eius anteriorem tua contingas lancea, pedem preponens sinistrum. Verum in ipso contactu observabis, num fortiter vel infirmiter lancea contineat. Si minus / fortiter restiterit, tum voluendo lanceam per manus retrorsum celeriter, introgreditor contra hostem dextro pede, et collum impulsu adpetas.

Sin vero is eodem habitu adversum te fuerit usus, sinistrum pedem praefigentem, manumque sinistram / iuxta genu laevum adposueris, posteriorem vero mucronem, iuxta pedem dextrum, intro ad hostem pede praenunciato concedas, et impulsum ipsius intra tuam manum utramque versus latus tuum dextrum avertere non dubitabis, interea autem dextrum pedem / reducito, atque lanceam in dextrum latus deflectes. Verum in ipso deflexionis motu per manus lanceam retrorsum si retorseris ad aequales certaminis labores ex eo habitu venire poteris mucrone anterioris porrecto, et interea lanceam ab eius / lancea si circumvolueris de hostis latere sinistro versus dextrum, visum ipsius pulsabis impulsus impetu.

Adversario autem simili contra te habitu utente, mucrone anterioris eum repellere memineris versus tuum latus dextrum, sed dextro pede reducto, / lanceae mucronem de latere eius dextro anteriorem contra hostis visum contorqueas, atque in tuta tuis corporis defensione ab eo recedas.

C93, 198v

Ain unnders einlauffen gögen ainnem unndern absetzen.

Item schickn dich also mit disem Zufechten wann dü dich zu dem mann fichst so pind im vornen an seinen spieß das / dein lincken fuoß vor stee unnd inn dem anpinden empfind ob er waich oder hert sey an seinem spieß ist er waich gegen dir unnd helt dir mit starckn wider so laß deinen spieß behend durch dein / hend zü rückn lauffen indes trit mit deinem rechten schenckel hinnein unnd stoß im nach seinem hals.

Ist er dir also einglauffen unnd stoß dir nach deinem hals unnd dü mit deinem lincken fuoß / vor steest dein lincke hannd bey deinem lincken knie den hindern ort bey deinem rechten schenckel so trit mit deinem rechten fuoß hinnein unnd seß im seinen ort ab zwischen deinen baiden / henden inn deinem spieß auf dein rechte seiten in dem seß deinen rechten schenckel wider zü rückn unnd wind dich mit deinem spieß auf dein rechte seiten unnd in dem winden laß deinen spieß zü rückn / lauffen durch deinen hend so kanstu mit deinem vordern ort mit Im zü gseicher arbait komen indes wechsel im durch an seinem spieß von siener lincken seiten auf sein rechte unnd stosß im damit nach / seinem gesicht.

Wechselt er dir also durch unnd stost dir nach deinem gesicht so seß im das ab mit deinem vordern ort auf dein rechten seiten unnd zeuch deinen rechte schenckel widerumb zü rückn / indes scheüß im deinen spieß aüf sein rechte seiten mit deinem vordern ort zü seinem gesicht unnd zeüch dich damit zü rückn inn güter versaßüng.

A low charge-in against a low displacement.

When you approach the opponent while leading with your left foot (as depicted by the left figure), bind him at the forward part of his longstaff and feel whether he is strong or weak. If he is weaker than you, then quickly pass your staff backward through your hands, step toward him with your right foot, and thrust into his throat.

If he thrusts toward your throat using the same technique while you are standing with your left foot forward, your left hand by your left knee, and your back point by your right foot (as depicted by the right figure), then step toward the enemy with your right foot, set his thrust to your right side between your hands, bring the right foot back, and wind the staff onto your right side so that you can work with the front point; then, change him through from his left side to his right and thrust violently into his eyes.

If he changes you through and thrusts at your face in this same way, then set him onto your right side with the front point, bring your right foot back, thrust your front point into his eyes from his right side, and withdraw from him in a strong guard.

LONGSTAFF 11

Cod. Vinob. 10825, 171r

Aversio contra impulsum liberum qui ex totis viribus procedit, ut lanceam contra visum hostis porrigas.

In hunc statum hac ratione te accomodabis, pedem sinistrum praepones, lancea supra caput elevata, manus dextra mucronis posterioris sit adplicata, sinistra autem versum mediam lanceae porrigatur, et mucro anterior terrae sit proclinatus. Et / si eveniat, ut adversarius ex habitu eo, quo lanceam in faciem tuum dirigat, sentiatque num acriter lancea resistas, te adgrediatur pedem sinistrum praefigens, impetum eius de dextro et sinistro latere repellas, atque lanceam inflectendo de latere tuo / dextro versus hostis itidem dextrum, anteriorem partem lanceae tolles sursum, atque dextro consequutus, lanceam in visum adversariis torqueas.

Sin vero idem ab eo tibi contigerit, sinistro retrorsum concedes, et manum dextrum si mutaris in / lancea, eaque conversa in latus tuum dextrum, eius impulsus avertitur. Verum per eam formam ex lancearum contactu, visum eius adgreditor versus latus dextrum adversariis. Caeterum si te repulerit avertendo, tum de latere ipsius dextro versus / sinistrum inflectendo lanceam pectori hostis infigito.

Cum vero is idem usurparit, te dextrum pedem praeponente, celeriter de latere eius sinistro dextrum versus circumvoluendo lanceam pudendis impellito, atque eo habitu retrorsum concedas, habita / tamen cura, ut tuto id fiat.

C93, 199r

Ein Abnemen gögen dem frey stoß so auß sterckne gat.

Item schickn dich allso Inn das abnemen gögen dem freystoß stee mit deinem linckn fuoß vor deinen spieß wol uber dein haüpt / die recht hannd hinden bey deinem Ort die linckn wol vornen In dem spieß das der vorder ort aüf der erdt ligt stost Er dann aüf dich aüß der sterckne seinen lincken fuoß furgeseßt so nimb Im das ab / von deiner rechten unnd lincken seitten unnd ganng vornen aüf mit deienm spieß Indes volg mit deinem rechten fuoß hinnach unnd stoß Im nach seinem gesicht.

Stost Er dir also zü deinem gesicht so / trit mit deinem lincken schenckel wider zü rückn unnd verwechsel die recht hand an deinem spieß unnd wend deinen spieß aüf dein rechten seiten so nimbstu Im seinen stoß damit ab In dem gang Im / aüß dem anpund zü seinem gesicht aüf sein rechte seiten seßst Er dir das ab so wind Im von seiner rechten aüf sein lincke seiten zü seiner prüst.

Stost Er dir also zü deiner prüst unnd dü mit deinem / rechten fuoß vorsteest so wechsel behend durch von seiner lincken aüf sein rechten seiten unnd stoß Im damit zü seiten gemechten Indes zeüch dich In dem wechsel zü rückn und hab deiner gesichts güte acht / Inn dem wind Im von deiner rechten aüf sein recht seiten.

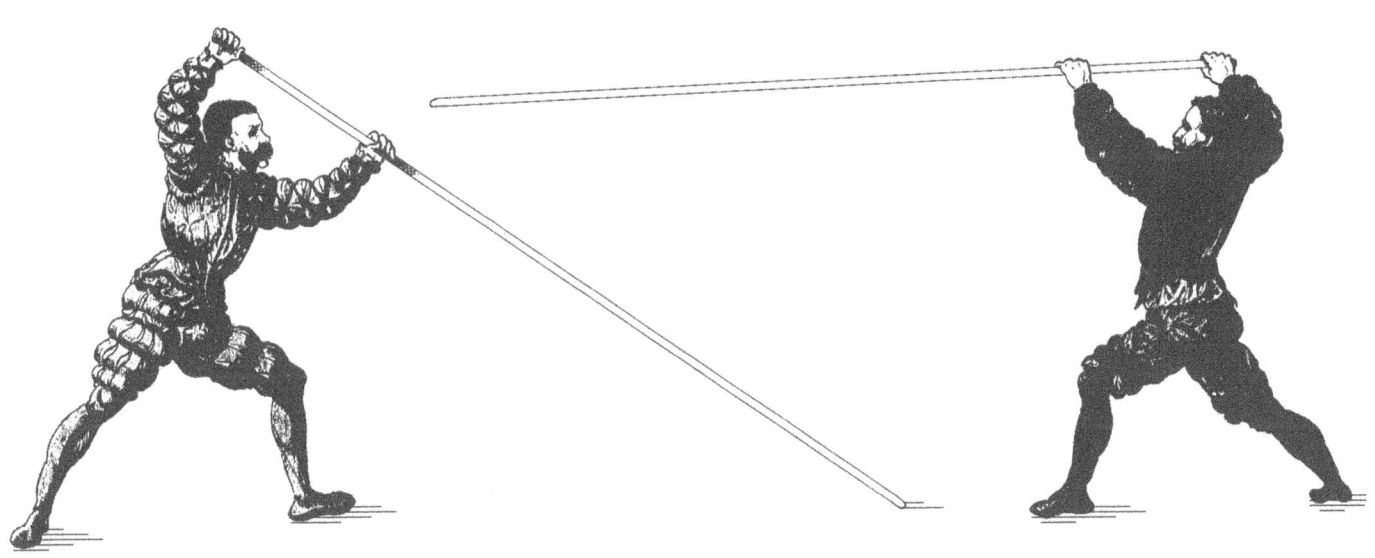

A parry followed by a thrust to the eyes against a free thrust.

Position yourself thusly against a free thrust: place your left foot forward and raise the longstaff above your head, with your right hand at the back point, your left hand extended in the middle, and your front point tilted toward the ground (as depicted by the left figure). If the enemy thrusts at your face from the strong with his left foot leading (as depicted by the right figure), then parry his attack on your right or left side, change him through from your right side to his left, lift your front point up, pass forward with your right foot, and thrust into his eyes.

If he thrusts at your face in the same way, then bring your left foot back, invert your right hand on the staff, wind it onto your right side, set his thrust aside, and thrust from the bind into his eyes on his right side. If he sets you aside, then wind him from his right side to his left and thrust into his chest.

If he thrusts thusly at your chest while you are standing with your right foot forward, then quickly change him through from his left to his right side, thrust into his groin, and withdraw backward, taking care to do so while guarded.

Longstaff 12

Cod. Vinob. 10825, 171v

Lancearum contactus, unde incursio procedas, qua hostis prosterni potest.

Quum commode hoc habitu uti voles, sinistrum pedem praepones, atque minus firmiter lanceam continge de latere hostis dextro, verum subito ea circumvoluta latus sinistrum impulsu adpetito.

Sin autem is eodem modo te adgressus fuerit, id animadvertentem tum lanceae medio eum propulsabis, dextro consequitor, atque manu utraque lanceam si sustuleris supra caput lanceam in visum vel pectus adversariis torqueto pro viribus.

Cum autem is idem sit conatus contra te, lancea abiecta, eius lanceam inflecte sursum de pectore tuo manu sinistra, interea vero confestim intro procedes, et lanceam hostis mediam adpraehendes.

Sed si tuum eadem ratione arripuerit adversarius, manum sinistrum ab lancea removebis, eaque locum, brachiis eius dextri secundum cubitum, infirmiorem adpraehendes, dextrum pedem utrisque eius pedibus postpones. Interim vero manu dextra locum debiliorem brachiis sinistri retro cubitum arripias, et si pro viribus brachia tua in formam crucis conposueris, ita ut interea caput tuum possis exerere. Eo igitur habitu si te converteris, quocumque volueris supra dorsum absque omnis tuis corporis detrimento apportare hostem licebit.

C93, 199v

Ein anpund darauß ain einlauffen geet mit ainem wurff.

Item schickn dich also inn den anpund stee mit deinem lincken fuoß vor unnd pind im mit der schwech an den spieß von seiner rechten / seiten indes wechsel im behend durch unnd stoß im seiner lincken seiten zü.

Stost er dir dann also nach deiner lincken seiten unnd dü desselben gewar wirst so nimb das mit deinem halben spieß ab volg mit deinem / rechten schennckel hinnach unnd far mit baiden henden mit deinem spieß wol aüf uber dein haüpt so hastu ainem vollen stoß aüf in unnd stoß im damit nach seinem gesicht oder der prüst.

Stost er also auß / der sterckn auf dich so laß deinen spieß fallen und wind im seinen spieß von unnden auf von deiner er prüst mit deiner lincken hand indes trit behend hinnein und erwisch im seinen spieß inn der mit.

Hat er / der deinem spieß also starckn erwist so laß dein lincke hand von deinem spieß unnd greif damit nach seiner schwech seines rechten arms hinder dem elnpogen trit mit deinem rechten schenckel hinder seine baid / schenckel inn dem greiff mit deiner rechten hand auf sein lincken arm inn die schwech hinder dem elnpogen unnd schrenckh dein arm mit gannßer sterkn das dü mit dem kopff zwischen deinen baiden armen / hindurch komest unnd wend dich damit so tregstu in aüf dem rücken wa dü in hin haben wilst unnd er mag dir sainen schaden thon.

A bind followed by a charge-in and a throw.

For this technique, place your left foot forward and bind with the enemy at the strong of your longstaff on his right side, then immediately change him through and thrust into his left flank.

If he thrusts to your left side in the same way and you anticipate it, then set him aside with the middle of your staff, pass forward with your right foot, lift the longstaff above your head with both hands, and thrust into his eyes or chest with all your might.

If attempts this same thrust against you, then drop your weapon, push his staff up away from your chest with your left hand while stepping quickly to the inside, and grab the middle of his lance.

If he catches your longstaff with this same technique, then release the staff with your left hand and seize the weak part of his right arm, behind the elbow; place your right leg behind both of his legs; grab the weak part of his left arm, behind the elbow, with your right hand; and cross your arms with all your might, until you can reach your head. From this position, if you turn yourself around, then you will be able to carry him on your back (as depicted in the illustration) and he will not be able to injure you.

Chapter 4

Halberd

"**BIPENNIS** antiquissimus usus, quod Amazonum inventum dicit, / continet, 20, optimos habitus, itidem ut in aliis factum, imaginibus depictos" (Cod. Icon. 393, 14r, 15-16).

"The **HALBERD** is an ancient skill, which is said [to be] an invention of the Amazons. [This chapter] contains the 20 best techniques, with images painted in the same manner as was done in the others." A technique from the "Various Weapons" chapter is also included here.

The halberd depicted in Mair's manual is essentially an ax mounted on a wooden haft, roughly six feet in length, with a spear point on the forward end and a hook opposite the blade. Note that, in some plates, the spear point is shown to have a rounded tip, clearly indicating a training weapon.

Although Mair traces the origin of the halberd to mythical Greece, it was, in reality, first developed in early 14th century France. Over the next two hundred years, the Swiss adopted it as their national weapon and further refined its use, incorporating halberdiers into their pike formations. However, by Mair's time, firearms had eroded the halberd's effectiveness as a battlefield weapon, and its design had become less streamlined and more ornamental than in previous centuries. The halberds shown here reflect this change.

Like the spear, the halberd may be used for thrusting, slashing, and striking, so much of the material from the preceding chapters applies. Additionally, the halberd's blades may be used to trap or hook the opponent.

Halberd 1

Cod. Vinob. 10825, 178r

Primi duo ictus superni bipennis de latere sinistro.

In hunc habitum hoc modo te compones, laevum pedem praeponito, bipennem supra caput contineas. Verum interea dextro pede consequutus / caput adversariis pulsabis.

Sed si is te ratione eadem fuerit adgressus, in habitu vicissim ictus supernis consistentem, sinistrumque praefigentem / laevum reducito retrorsum, et si pariter cum adversario ferias, eius ictus irritus fiet, interim vero bipenne sublata, faciem / hostis impulsu adpetes anterioris mucronis. Sin is exceperit, ea rotata de latere eius sinistro versus dextrum, dextro eius brachio / bipennis laminam adplicabis.

Caeterum si is idem usurparit contra te, dextro pede si recesseris, feriundo ab hoste te liberaveris, sed mox / bipennem sursum vibrabis coram ipsius conspectu atque in ipso vibrationis habitu, eius bipennis adplicabis tuam de latere eius sinistro / verum ex ea forma bipennum si deflexeris, versum te adtrahes. Porro si hostem firmiter resistentem senseris, rursum dextro consequutus / pectus adversariis pungito.

Ipso autem eo habitu utente, rursus laevo reducto, mucrone posterioris impulsum repellito, et hac / ratione te ab omnis periculo vindicabis.

C93, 202r

Die ersten zwen oberhew inn der hellenparten vonn der lincken seiten.

Item schickn disem stuckn dich also mit dem zfechten stannd / mit deinem lincknen füoß vor unnd halt dein hellenparten ob deinem haüpt indes volg mit deinem / rechten füoß hinnach unnd haw im nach seinem haüpt.

Hawt er dir dann also nach deinem / haüpt unnd dü auch gögen im stesst inn dem oberhaw mit deinem lincken füoß vor so seß den / zü rückn unnd haw auch von oben gleich mit im ein so ist sein haw umb sunst inn dem gang aüff / unnd stoß im zü seinem gesicht mit deinem vordern orth verseßt er dir das so wechsel durch von / seiner lincken aüf sein rechte seiten unnd seß im mit deinem plat an seinen rechten arm.

Hat er dir / also angeseßt so trit mit deinem rechten schenckel zü rückn so hawstü dich damit von im inn dem / streich auf mit deiner hellenparten im vor seinem gesicht unnd in dem streichen pind im an sein / hellenparten aüf sein lincke seiten indes wend dein hellenparten unnd Raiß damit zü dir ist er / starckn unnd wilt dir mit nach hengen so trit mit deinem rechten schenckel wider hinein und stoß / in zü seiner prüst.

Stost er dir also zü deiner prüst so trit mit deinem lincken schenckel wider yü rückn / unnd seß im den stoß ab mit deinem hindern orth so bistü sein on schaden ledig.

The first two *oberhawen*[1] with the halberd from the left side.

Position yourself thusly for this technique: place your left foot forward and hold the halberd above your head (as depicted by the left figure), then pass forward in that same instant with your right foot, and strike the enemy's head.

If he strikes at your head with this same technique while you are standing in the guard for an *oberhaw* with your left foot leading (as depicted by the right figure), then bring your left foot back, strike at the same time he does, so that his own strike is ineffective, and thrust your front point into his face. If he sets you aside, then change your halberd through from his left side to his right, and press your blade to his right arm.

If he has you thus confined, then bring your right foot back, strike your halberd up into his face, hook his halberd on his left side, turn your blade, and pull him toward you. If he is strong, then pass forward again with your right foot and thrust into his chest.

If he thrusts at your chest in this same way, then bring your left foot back again and set him aside with your back point.

---

1. German, plural of *oberhaw*, "over strike; strike from the upper openings." See Latin *ictus superus*, "above strike."

Halberd 2

Cod. Vinob. 10825, 178v

Habitus inferi ictus bipennis de latere utroque.

Ad habitum ictus inferi ea, quae mox sequetur, ratione te accomodato, pedem sinistrum praepones, deorsum mucro versus adversarium / vergat, et confestim manum dextram in coxam dextram dirige, atque bipennem versus hostis visum vel pectus propellito. Sin is averterit / tuum impulsum, tum ab eius sinistro in latus dextrum bipennem rotata.

Verum si gemino impulsu is contra tuam faciem fuerit usus / pedem sinistrum te praefigente itidem in habitu ictus inferi, tum manu dextra bipennem continente femoris dextro adplica, atque lamina / de latere tuo dextro adversariis impulsum refringes, verum pede consequutus dextro, pectori eius bipennis mucronem anteriorem / infigito.

Sed si idem contra te hostis usurparit, sinistrum pedem retrorsum diriges, atque mox dextro itidem reducto, ad / aequales conatus certaminaque cum hoste perveneris. Sed confestim ictu supero caput eius conquassato.

Si autem is superne te / fuerit adgressus, bipenne celeriter sursum sublata, eius ictum lamina tuae bipennis excipies, et si pudenda eius impulsu adpetiveris / retrorsum concedas licebit.

C93, 202v

Der unnderhaw in der hellparten von baiden seiten.

Item schickn dich also mit disem zufechten in / den unnderhaw stee mit deinem lincken fuoß vor unnd halt den orth undersich gögen dem man in / dem ganng mit deiner rechten hand aüf dein rechte hufft unnd stoß im damit zü seinem gesicht oder / der prüst seßt er dir den stoß ab so wechsel durch von seiner lincken auf sein rechte seiten.

Stost er dir / also zwifach zu deinem gesicht unnd dü mit deinem lincken fuoß vorsteest auch in dem underhaw / so far mit deiner rechten hand in dein rechte huff mit deiner hellenparten unnd seß im das ab aüf / dein rechten seiten mit deinem plat indes volg mit deinem rechten schenckel hinnach und seß im deinen / vordern orth an sein prüst.

Seßt er dir den ort also an dein prüst so trit mit deinem lincken / schenckel zü rückn indes zeüch den rechten aüch behend zü rückn so magstü wider mit inn zü gleicher / arbait kummen unnd haw im mit ainem oberhaw zü seinem haüpt.

Hawt er dir also oben ein / so far aüf mit deinerm hellenparten von unnder so fach stü seinen haw inn dein plat indes zückn / und stoß im nach seinen gemechten unnd zeüch damit zü rückn.

An *underhaw*[1] with the halberd from both sides.

Position yourself thusly for this *underhaw*: place your left foot forward and tilt your point down toward the adversary (as depicted by the left figure), then immediately bring your right hand to your right hip, and thrust into his face or chest. If he sets your thrust aside, then change him through from his left to his right side.

If he thrusts twice at your face in this way while you are standing in an *underhaw* with your left foot leading (as depicted by the right figure), then bring your right hand to your right hip, set his thrust onto your right side with the blade, immediately pass forward with your right foot, and thrust your front point into his chest.

If he thrusts his point at your chest in the same manner, then step back with your left foot, immediately follow with your right foot so that you have room to maneuver, and strike an *oberhaw*[2] to his head.

If he strikes an *oberhaw* at you, then quickly raise your halberd up high, parry his blow with your blade, thrust into his groin, and withdraw from him.

---

1. German, "under strike; strike from the lower openings." See Latin *ictus infernus*, "lower strike."
2. German, "over strike; strike from the upper openings." See Latin *ictus superus*, "above strike."

Halberd 3

Cod. Vinob. 10825, 179r

Mutatorius cancellatus contra habitum aversionis

Quum rite habitu modo praedicto uti volueris, pedem sinistrum praepones, manus sinistra sub axilla dextra collocetur, et ex ea / forma dextro introgreditor, inflectes dupliciter in ictum inferum, atque pedem adversariis sinistrum impulsu quassabis.

Sed / si idem ab hoste fieri animadverteris contra te in habitu aversionis consistentem, sinistrumque pedem praefigentem, tum si sursum manum / dextram rexeris, impulsus eius ad irritum redigetur, et confestim dextro pede consequutus, mucronem anteriorem hostis vultus / infiges.

Cum vero tuum adpetierit visum, eum avertere non dubitabis ex latere tuo sinistro versus dextrum loco firmioris, interim / ab utroque latere vibrabis bipennem in hostis conspectu. Verum ex ipsa vibratione, eius bipennem si contigeris, dupliciterque ea / rotata visum eius impulsu adgreditor.

Si autem idem is contra te usurparit, lamina tuae bipennis avertes ita, ut eius / impulsus supra caput tuum ex ea repellendo. Caeterum ex eo habitu sinistro pede insequitor, et si mucrone posterioris adversariis / bipennem repuleris, liberum te atque securum reddideris.

C93, 203r

Der geschrencknt wechsel gögen ainem abnemen.

Item schickn dich also in den geschrennctien wechsel stee / mit deienm lincken fuoß vor dein lincke hand under deiner reichten achsen indes trit mit deinem / rechten schenckel hinnein wind dich damit zwifach in den unnderhaw unnd stoß im nach seinem / lincknen schenckel.

Wirstu des stoß gewar unnd steest gögen im inn dem abnemen deinem lincken fuoß / furgeseßt so far mit diener reichten hannd auf deie rechte seiten so nimbstü im seinen stoß hinweckn / indes volg mit deinem rechten schennckel hinnach unnd stoß im deinen vordern ort zü seinem gesicht.

Stost er dir dann also zü deinem gesicht so seß im das ab von deiner lincken auf dein rechten seiten / mit mit der sterckn indes streich auf von ainer seiten yü der anndern im vor seinem gesicht und inn / dem auf streichen pind im an sein hellenparten inn dem wechsel im zwifach durch unnd stoß im nach / seinem gesicht.

Sotst er dir also zu so nimb im das ab mit deinem plat das dir stoß ob deinem haüpt / hinauß ganng in dem volg mit deinem lincken schenckel hinnach unnd schlag mit deinem hin / dem ort sein hellenparten hinweckn so bistü von im ledig.

The crossed *wechsel*[1] against an *abnemen*.[2]

Position yourself thusly in the crossed *wechsel*: stand with your left foot forward and your left hand gripping the haft underneath your right arm (as depicted by the left figure), then step to the inside with your right foot, wind twice into an *underhaw*[3] (as depicted by the right figure), and thrust into the adversary's left foot.

If you anticipate this same thrust from the enemy while you are standing in the *abnemen* with your left foot leading, then bring your right hand up so that his thrust is blocked, immediately pass forward with your right foot, and thrust your forward point into his face.

If he thrusts at your face in this way, then set him from your left side to your right with the strong and immediately strike into his face, from one side to the other. If you bind his halberd in the process, then change him through twice and thrust into his eyes.

If he thrusts thusly toward you, then set him aside with your blade in such a way that his thrust is deflected over your head, then pass forward with your left foot and push his halberd away with your rear point, so that you are free from him.

---

1. German, "changer." See Latin *mutatorius*, "change."
2. German, "taking-off." See Latin *habitus aversionis*, "parry; setting-aside."
3. German, "under strike; strike from the lower openings." See Latin *ictus infernus*, "lower strike."

Halberd 4

Cod. Vinob. 10825, 179v

Impulsus quo visus adpetitur contra eum, qui de pectore formatur, et pectus itidem eius sauciatur.

Ad eum habitum hac ratione te compones, sinistrum pedem praefigito, bipennem iuxta thoracem contineto ita, ut acies superne sic conversa, / et consequutus dextro visum adversariis adgreditor.

Sin autem adversarius idem adversum te usurparit, sinistrum pedem / praeponentem, eum sursum ictum qui de thorace formatur dirigas, atque eius impulsum hac ratione avertito. Verum celeriter dextro / pede consequutus, pectus eius impulsu concuties. Sed si hostis eum repellendo averterit, tum bipennem tuam ab eius circumvoluas / de eius latere dextro versus sinistrum, inde autem pedem sinistrum eius dextro si praeposueris, gemino impulsu hostis visum adpetito.

Quum vero idem contra te usurparit, ab utroque latere te defendes eius impulsum repellendo. Pede sinistro intro procedas / bipenne rotata de latere adversariis dextro in sinistrum, laevum brachium eius quassato impulsu. Eo impetum excipiente, tum bipennis / mucronem posteriorem sursum de terra levando reges versus hostis pectus.

Sed si is habitu praedicto contra te fuerit usus, sinistro / pede reducto eius inflexio impedietur, subito interim adversariis vultum impulsu sauciabis, atque retrorsum concedas.

C93, 203v

Ein gesichtstoß gögen dem prusthaw.

Item schickn dich also in den gesicht stoß stannd / mit deinnem lincken fuoß vor unnd haltt dein hellenparten aüf der prust das die schnied oben gewendt / sey indes volg mit deinem rechten schenckel hinnach und stoß im zü seinem gesicht.

Stost er dir / also zü deinem gesicth unnd dü mit deinem lincken fuoß vor stest so gee aüf mit deinem prusthaw / unnd seß im seinen stoß damit ab in dem so trit mit deinem rechten schenckel hinnein und stoß im / zü seinem gesichte.

Stost er dir also zwifach zü so seß im den stoß ab von baiden seiten unnd trit mit / deinem lincken schenckel hinein unnd wechsel im durch von seiner rechten aüf sein lincken seiten / unnd stoß im damit zü seinem lincken arm verseßst er dir den stoß so wind im deinen hindern / ort von unnden aüf zü seiner prüst.

Windt er dir also seinen hindern ort zü deiner prüst so trit / mit deinem lincken schenckel zü rückn so ist sein winden umb sunst in dem stoß im zü seinem gesicht / und zeüch dich damit zü rückn mit gannzer unnd halber hellenparten.

A thrust to the face against a *prusthaw*[1]

Position yourself thusly for a thrust to the face: stand with your left foot forward and hold the halberd near your chest in such a way that your blade is turned upward (as depicted by the left figure), then pass forward with your right foot and thrust into the enemy's face.

If he attempts this same thrust to your face while your left leg is leading, then strike a *prusthaw* upward, set his thrust aside, then quickly pass forward with your right foot and thrust into his face or chest.[2]

But if he sets your thrust aside, then wind your halberd around his, from his right side toward his left, place your left foot in front of his right foot, and thrust twice into his eyes.[3]

If he thrusts twice toward you in this way, then parry him on both sides, step to the inside with your left foot, change him through from his right side to his left, and thrust into his left arm. If he parries your thrust, then bring your back point up from below and wind it into his chest.

If he winds his back point to your chest thusly, then step back with your left foot, thrust into his face, and withdraw backward with a full or half-halberd guard.

---

1. German, "breast strike." See Latin *ictus pectoris*.
2. In the Latin, the target is the enemy's *pectus*, "chest"; in the German, it is his *gesicht*, "face."
3. This sentence does not appear in the Dresden MS.

Halberd 5

Cod. Vinob. 10825, 180r

Bipennium contactus vel collisio unde habitus vellendi procedit.

In praedictum modum ita te adaptabis, sinistrum pedem praeponito, bipennis mucro posterior femoris dextro sit adplicatus, anterior / versum adversariis visum directus atque ex latere tuo dextro, in hostis sinistrum, eius bipennem contingito, interimque tuam bipennem celeriter supra eius inflectito, et ad te adtrahito ex eo habitu.

Sin autem ita contra adversarium constiteris in / habitu contactus bipennium sinistrum pedem praeponens, isque ad se vulserit, tum consequitor dextro, atque versum pectus super / ipsius brachium pungito. Sin vero impulsum animadverterit adversarius, simulque avertere conatus est, bipenne rotata seu / circumvoluta de latere hostis sinistro in latus eius dextrum versus visum adversariis bipennem propellito.

Sed si aliam nuditatem / ex alia tibi requisiverit, sinistrum pedem reducito, atque eius conatum ex utroque latere mucrone posterioris. Hoc itaque perfecto, / rursum intro pede concedas sinistro, atque lamina bipennis porrecta, eius bipennem eo habitu repellas, subito autem eum in visum / hostis coniscito. Caeterum si is id averterit, aliam ex alia nuditatem si conquisiveris in transmutatione, dupliciter ab / adversario recedere memento.

C93, 204r

Ein anpund darauß ain reyssen geet.

Item schickn dich allso in dem anpund stand mit deinem / lincken fuoß vor dein hellenparten mit deinem hindern orth aüf deiner rechten hüff den orth / gögen seinem gesicht unnd pind im an vor deiner rechten seiten aüf sein linke inn dem wind dein / hellenparten behend ober die sein unnd reisz damit zü dir.

Statu dann auch allso gögen im inn / dein anpund mit deinem lincken fuoß vor unnd er also an sich reist so volg mit deinem rechten / schenkel hinnach unnd stoß im zu seiner prust hinnein ob seinem lincken arm wirt er das stoß / gewar unnd verseßt dir den so wechsel durch von seiner lincken aüf sein rechten seiten unnd stoß im / zu seinem gesicht.

Sucht er dir deine plossen also von ainer seiten zu der anndern so trit mit deinem / lincken fuoß zu ruckn unnd seß im das ab mit deinem hindern ort auf baid seiten in dem / trit mit deinem lincken schenckel wider hinnein unnd wind dich mit dem plat hin fur und nimb / im damit sein hellenparten hinweckn indes stoß im zu seinem gesicht seßt er dir das ab so such im / seine plossen von ainer seiten zu der andern in dem durch wechsel unnd zeuch damit zwifach ab..

A bind followed by a pull.

Position yourself thusly for this technique: place your left foot in front, with your rear point at your right hip and your forward point aimed at the adversary's eyes, then bind him from your right side to his left, quickly wind your halberd over his (as depicted by the left figure), and pull toward you.

If you are standing thus against him in the bind of the halberds with your left foot leading and he pulls you toward him, then pass forward with your right foot, and thrust over his arm, into his chest.

But if he anticipates your thrust and attempts to set you aside, then change him through, from his left side to his right, and thrust into his eyes.

If he seeks your openings from one side to the other, then bring your left foot back and parry him on both sides with your rear point, then step to the inside again with your left foot, bring your blade forward, push his halberd away, and immediately thrust into his face. If he sets you aside, then seek out his openings by changing through from one side to the other, and step away from him twice.

Halberd 6

Cod. Vinob. 10825, 180v

Bipennium contactus addito habitu quo adversarius intercludi potest superne.

Quum in eam formam in congressu te apte accomodare voles, sinistrum pedem praefigere memineris. Et si is vicissim contra te / in habitu iam praedicto sinistrum itidem praeponens constiterit, partem anteriorem hostis bipennis contingito. Verum in ipso / contactu, laminam tuae bipennis supra eius mucronem recurvum, qui e regione laminae bipennis affixus est, inflectes.

Sed si is usus fuerit / eadem ratione, bipennem converte, eaque sursum levata, celeriter superne eam arripe, atque visum vel pectus adversariis / figito. Sin autem eum impulsum adversarius repulerit, tum bipennem coram ipsius conspectu strenue vibrabis, interimque nuditatem / eius proximam requirito.

Sed si is vicissim tuam eodem modo perquisiverit, mucrone anterioris eum repellas licebit, dextro / confestim insequitor, eius bipennem mucrone posterioris si averteris, dupliciter ab loco infero sursum inflectendo versus / faciem vel pectus adversariis bipennem impellito. Sin autem is id animadverterit, et retrorsum concesserit, dupliciter ictibus / et inflexionibus instando hostem consequitor.

Sin autem is eadem usurparit, tum mucrone bipennis tuae anteriore pariter et posteriori eum repelles, inde retrorsum concedes habitu tuis visus cura diligenti.

C93, 204v

Ein anpinden mit ainem obern sperzen.

Item schickn dich also mit disem zufechten stannd mit / deinen lincken fuoß vor stat er dann auch gögen dir inn gleichem anpund seinen lincken fuoß furgeßt / so pind im vornen an sein hellenparten unnd in dem anpund so wind im dein plat ober seinen / hagcken.

Begert er dir also ein zuwinden so wend dein helleparten unnd schuch dann ober sich unnd in / dem zuech dein hellenparten oben behend unnd stost im zu seienem gesicht oder der prust seßt er dir / den stoß ab so streich im mit deiner hellenparten vor seinem gesicht auf in dem such im du nechft plose.

Sucht er dir deine plossen allso so seß im das ab mit deinem vordern ort deiner hellenparten volg mit / deinem rechten fuoß hinnach unnd schlag im mit deinem hindern orth sein hellenparten hinweckn / unnd wind im damit zwifach von unndern zu seinem gesicht oder der prust wirt er des ein windens / gewar unnd trit damit zu ruckn so raiß im zwifach nach mit hewen unnd mit winden.

Raist er dir / also mit disem stuckn nach so seß das mit deinem hindern und vordern orth deiner hellenparten / trit damit zu ruckn unnd hab deiner gesichts acht mit guter versaßung.

A bind with an upper block.

Position yourself thusly in the bind: place your left foot forward, and if your opponent is also leading with the left foot, then bind him on the front part of his halberd and wind your blade over his hook (as depicted by the left figure).

If he winds you with the same technique, then turn your halberd, raise it up high, quickly draw it to you, and thrust into his face or chest. If he sets your thrust aside, then flourish your halberd in front of his face and seek out his next opening.

If he seeks out your openings in the same way, then set him aside with your front point, immediately pass forward with your right foot, set his halberd aside with your back point, wind him twice from his lower openings upward, and thrust into his face or chest. If he anticipates this and steps back, then press him hard by striking and winding twice.

If he employs the same technique, then parry him with both your front and back points, and withdraw in a strong guard.

Halberd 7

Cod. Vinob. 10825, 181r

Ratio, qua contra adversarium superne ferire licet, pariter addito habitu inferne vellendi.

In hunc modum ita te adaptabis, sinistro pede praefixo in habitu ictus superi bipennem tuam contineas. Et si is vicissim contra te / sinistrum praeposuerit, porrecta bipenne sua contra faciem tuam, tum ictu supero eum repellas. Interea autem versum te velles de latere / tuo sinistro in dextrum, ex eo igitur habitu bipennem contra hostis visum dirigas.

Sed si is idem usurparit, laminam bipennis / tuae retrorsum eius pedi adiunges, et si versum te adtraxeris, conatus eius irritus fiet, simulque hostem ea ratione prosternere / poteris. Confestim etiam instando consequutus visum eius impulsu conquassabis.

Verum si is eadem ratione tibi institerit, / bipenne media eum avertes, eaque rotata de alio latere in aliud, proximam nuditatem hostis conquires. Cum vero se ab eo discrimine / se liberum reddiderit, tum pariter eodem momento eum ipso ferias. Sed in ipso bipennem contactum mucrone posterioris, eius / anteriorem deflectes, pedem dextrum interius hostis sinistro praepones, interim vero mucronem posteriorem circa eius collum / ex latere dextro adplicabis, et si eo habitu adversarium adtraxeris, facile eum prosternes.

Sin autem te prosternere conabitur, / manum utramque commutabis, atque tuam bipennem dextro eius brachio applicato, ea ratione si eum propuleris, te liberabis.

C93, 205r

Ain obers einhawen mit ainem undern Reyssen.

Item schickn dich Allso inn das ober einhawen stee mit deinem / lincken fuoß vor unnd halt dein hellenparten Inn dein oberhaw stat er dann auch gögen dir seißnen / linken fuoß fur geseßt sein hellenparten dir gögen deinem gesicht so nimb Im das ab mit dein ober / haw indes raiß so zu dir von deiner lincken auf dein rechten seiten unnd In dem reissen gang auf mit / deiner hellenparten im zu seinem gesicht.

Begert er dir also zu deinem gesicht zu geen so wend dein hellenparte / mit dem plat hinder seinen schenckel unnd zeuch damit an dich so Ist sein stoß umb sunst und / magst im auch damit fellen In dem raiß im nach mit ainem stoß zu seinem gesicht.

Stost er dir also zu / deinem gesicht so numb Im das mit deiner halben hellneparten ab unnd wechsel Im durch von ainer / seiten zu der anndern unnd such Im damit die nechsten plossen seßt er dir das ab so pind gliech mit Im / an sein hellenparten unnd Inn dem anpund wind mit deinem hindern Ort seinen vordern hinweckn / unnd trit mit deinem recht fuoß Inn wendig fur seinen lincken Indes far im mit deinem hinder orth / umb seinen hals auf sein rechten seiten zeuch In damit zu dir so magstu In werffen.

Begert er dich also zu / werffen so ver wechsel baid hendan deiner hellenparten unnd seß Im dein hellenparten ab seinen rechten / arm scheut In damit von dir so bistu seinen schaden ledig.

An inside *oberhaw*[1] with a lower pull.

Position yourself thusly for this technique: stand with your left foot forward and hold your halberd in the guard for an *oberhaw*. If he also stands with his left foot leading, and his halberd is aimed at your face, then set him aside with an *oberhaw*, pull his weapon from your left to your right side (as depicted by the left figure), and then thrust your halberd into his eyes.

If he employs this same technique, then hook your blade behind his foot, pull it toward you so that he cannot thrust into your face, throw him, and then thrust into his eyes.

If he thrusts thusly to your face, then parry him with the middle of the halberd, change him through from one side to the other, and seek out his nearest opening. If he sets you aside, then strike and bind with him, wind his front point away with your back point, place your right foot on the inside of his left foot while hooking your back point around his neck from the right side, and pull him toward you to throw him down.

If he tries to throw you, then switch your hands on the weapon, press your halberd to his right arm, and free yourself by pushing him off.

---

1. German, "over strike; strike from the upper openings." See Latin *ictus superus*, "above strike."

Halberd 8

Cod. Vinob. 10825, 181v

Habitus super intorsionis, quo hostis bipennis deorsum deprimitur ex contactu primo, addito habitu infere interclusionis.

Ad eum habitum hac ratione te compones, pedem sinistrum praeponito, bipennem supra pectus de latere dextro contineas, mucro in adversariis / visum sit directus, ex eo habitu dextro intro procedas, et ictu converso, ictus autem is conversus est, si bipennem contineas / iuxta parte anteriorem, dextra vero posteriorem atque is circumagatur sub brachio sinistro, ut anterioris parte caput eius / pulses caput eius quassato.

Sin autem te superne adgressus fuerit pedem sinistrum praeponentem, bipennis lamina eum repelles, / atque eodem momento cum hoste ferias. Verum in ipso bipennum contactu, ea rotata de latere eius sinistro versus dextrum, / trans hostis bipennem, tuam superinflecte, et si firmiter deorsum suppresseris, intercludetur adversarius.

Cum vero is / idem usurparit, sinistro consequutus pede, sursum firmiter bipennem tuam inflectendo attolles, eo tamen pacto, ut mucro / versus adversariis visum sit porrectus. Hoste id excipiente, dextrum pedem eius sinistro praepones, atque inter eius brachium / utrumque mucronem posteriorem inferere memineris supra brachium dextrum. Ex eo igitur habitu si depresseris de latere / tuo dextro, caput eius tuae bipennis lamina ferire poteris.

Caeterum si hostis idem perficere conatus sit adversus te, / tum manum sinistrum bipennis applicatam rursus transmutabis, sinistrum pedem si reduxeris, caput hostis mucrone posteriori / sauciato.

C93, 205v

Ein oberwinden auf dem anpund mit ainem unndern sperzen.

Item schickn dich also In diseß stuckn stannd mit deinem / lincken fuoß vor dein hellenparten auf deiner rechten prust den ort gögen seinem gesicht Indes / trit mit deinem rechten fuoß hinnein unnd haw Im mit ainem verkerten haw nach seinem haupt.

Hawt Er dir also oben ein unnd du mit deinem lincken fuoß vor stast so verseß Im das mit deinem / plat Inn dem pind gleich mit Im an unnd In dem anpund wechsel Im durch von seiner lincken / aüf sein reichten seiten ober wind Im damit sein hellenparten unnd truckn starckn unnder sich.

Hat er / dich also gespert so volg mit deinem lincken fuoß hinnach unnd wind dich mit deiner hellenparten / widerumb starckn aüf Im den Ort zu seinem gesicht seßt er dir das ab so trit mit deinen rechten / fuoß für seinen lincken unnd wind dich mit deinem hindern ort zwischen seinen baiden armen hinein / ober seinen rechten arm truckn damit unnder sich aüf dein rechten seiten so magstu In mit deinem / plat zu seinem haupt schlagen.

Hat er dir also hert ein gewunden so wechsel dein lincken hand / behend winderumb durch an deiner hellenparten Indes trit mit deinem lincken fuoß zu ruckn / unnd schlag Im mit deinem hindern ort nach seinem haupt.

Halberd

A high wind from the bind, with a low block.

Position yourself thusly for this technique: place your left foot forward and hold your halberd above chest-level on your right side, with your left hand forward, your right hand at the back point, and your front point aimed at the adversary's eyes; then, step to the inside with your right foot and strike his head with an inverted strike.

If he strikes you from above while your left foot is leading, then displace him with your blade, bind with him as he strikes, change him through from his left side to his right, wind over his halberd, and press his weapon down strongly.

If he presses your weapon down thusly, then pass forward with your left foot, strongly wind your halberd upward, and thrust your point into his face. If he sets you aside, then place your right foot in front of his left, wind your back point between both of his arms, over his right arm, and push him down onto your right side; then, you should strike his head with your blade.

If he attempts to do the same thing to you, then quickly change your left hand to an inverted grip, bring your left foot back, and strike his head with your back point.

Halberd 9

Cod. Vinob. 10825, 182r

Interior impulsus, quo brachium adpetitur contra visus impulsum.

    Si ad invicem concesseritis, et adversarius visum fuerit impulsu adgressus tuum, pede sinistro intro procedas, et interius brachium / eius sinistrum impulsu quassato, et confestim bipennis lamina eidem adplicata, in latus eius dextrum ea ratione hostem convertes, / eiusque impulsus ad nihilum redigitur, inde vero bipennem retrahito, atque pectus hostis mucrone anterioris adpetito.

    Sin vero is pectori / tuo eodem modo immineat. Sinistrum pedem reducere non dubites, eumque lamina, id est parte bipennis planida avertes, postea bipennem / in eius conspectum tolles, et rursum sinistro introgressus, mucronem vultus eius vel pectoris infigito.

    Sed si hostis idem usurparit, / defendendo te de alio latere in aliud eum avertes, atque ex ea forma dextro pede intro si processeris, mucrone posteriori caput eius / pulsato. Excipiente id adversario, tum pede sinistro in latus hostis dextrum si concesseris, lamina tuae bipennis brachium eius dextrum / quassabis.

    Verum si is eodem habitu contra te fuerit usus, mucrone eum posterioris repellas, et anteriorem inferne versus pectus eius / impellito. Postea dextro pede introgressus, caput adversariis mucrone posteriori pulsato.

C93, 206r

Ain Inwendiger arm stoß gogen ainem gesicht stoß.

    Item wann Ir mit dem zufechten zu sainen komen unnd er dir zu / dem gesicht stost so trit mit deinem lincken fuoß hinnein unnd stoß im in wendig nach seinem lincken / arm unnd in dem seß im das plat an so scheußtu in damit von dir aüf sein rechten seiten unnd ist sein / stoß umb sunst indes ruckn dein hellenparten behend zu dir unnd stoß im mit deinem vordern ort zu / seiner prust.

    Stost er dir also zu deiner prust so seß denen lincken schenckel zu ruckn und seß im das ab / mit deinem plat inn dem gannng mit deiner hellenparten aüf im fur sein gesicht trit mit deinem lincken / fuoß widerumb hinnein unnd wind im deinen ort zu seinem gesicht oder der prust.

    Windt er dir / also deinem gesicht zu so nimb im das ab von ainer seiten zu de anndern inn dem trit mit deinem rechten / fuoß hinnein unnd schlag im mit deinem hindern ort zu seinem haupt verseßt er dir den schlag so trit / mit deinem lincken fuoß aüf sein rechte seiten unnd haw im mit deinem plat nach seinen rechten arm.

    Hawt er dir also zu so nimb im das ab mit deinem hindern ort unnd wind im deinen vordern von unden / ein zu seiner prust indes trit mit deinem rechten schennckel hinnein unnd sclag im mit deinem hindern / ort zu seinem haupt.

An inside thrust to the arm against a thrust to the face.

If the opponent thrusts at your face as you approach, then step to the inside with your left foot, quickly thrust into his left arm, press your blade against it (as depicted by the left figure), turn him onto his right side so that his attack is useless, pull your halberd back, and thrust your front point into his chest.

If he thrusts thusly at your chest, then bring your left foot back, set him aside with your blade, bring your halberd into his face, step to the inside again with your left foot, and wind your point into his face or chest.

If he winds thusly to your face, then set him from one side to the other, simultaneously step to the inside with your right foot, and strike his head with your back point. If he parries this, then step to his right side with your left foot and strike your blade into his right arm.

If he strikes you in this way, then set him aside with your back point, wind your front point underneath to thrust into his chest, advance with your right foot, and strike your rear point into his head.

Halberd 10

Cod. Vinob. 10825, 182v

Pectoris impulsus addito simul habitu interclusionis.

Quum rite et more Athletarum strenuorum praedicto habitu uti volueris, sinistrum pedem praeponas, supra partem pectoris dextrum, bipennem / contineas, et mucrone anteriori pectus adversariis impulsu adpetas.

Sin autem contra hostem constiteris dextru praefigens / teque eadem ratione adpeti[v]erit, tum parte bipennis tuae anterioris eum repelles, subito autem tuum bipennem diriges supra adversariis / bipennis laminam, eoque habitu te in latus tuum sinistrum si converteris, eius bipennem intercludes. Verum in ipsa interclusione, sinistram manum / dextrae propius admove, et dextrum pedem si retuleris, versus caput eius ferias.

Sin autem is idem usurparit contra te, sinistrum reducas, atque / mucrone anteriori ictum ipsius avertito, inde autem firmiter eum sursum adtolles, intro dextro pede concedas, et mucronem posteriorem tuae / bipennis collo eius adplicato. Caeterum si id hostis averterit, sinistro celeriter consequutus, laminam bipennis capiti eius infligito.

Sin autem / adversarius habitu eodem usus fuerit, tum intra manum utramque excipito eius impetum in latus tuum dextrum, dupliciterque retrorsum concedas / feriundo.

C93, 206v

Ain prust stoß mit ainnem sperzen.

Item schickn dich also in diseß stuckn stand mit deinem lincken / fuoß vor unnd halt dein hellemparten ob deiner rechten prust indes stoß im mit deinem vordern orth / deiner hellenparten nach seiner prust.

Stastu dann auch also gögen im mit deinem rechten fuoß vor / unnd er dir also zu stost so seß im den stoß ab vornen mit deiner hallenparten indes oberfall im / mit deiner hellenpartem uber sein plat unnd verter dich damit auf dein lincke seiten so sperystu im / sein hellenparten unnd inn dem speryen greif mit deiner lincken hand zu deiner rechten in dem seß / deinem rechten fuoß zu ruckn unnd haw im oben nach seinem haupt.

Hawt er dir also oben zu so / trit mit deinem lincken schenckel zu ruckn unnd seß im das ab mit deinem vordern ort deiner hellenparten / in dem scheuß in starckn mit deiner hellenparten ober sich trit mit deinem rechten fuoß hinnein / und wind im deinen hindern ort zu seinem hals seßt er dir das ab so volg mit deinem lincken fuoß / hinnach unnd haw im mit deinem plat zu seinem haupt.

Hawt er dir also oben ein so verseß im das / zwischen deinen baiden henden in dem hellenparten auf dein rechte seiten haw dich damit zwifach / zu ruckn.

A thrust to the chest with a block.

Position yourself thusly for this technique: stand with your left foot forward, hold the halberd above your right breast, and thrust your front point into the adversary's chest (as depicted by the left figure).

If you are standing with your right foot leading and he thrusts at you in the same manner, then set his thrust aside with the forward part of your halberd, bring your weapon over his blade, and turn to your left side so that you block his halberd (as depicted by the right figure), then move your left hand closer to your right hand, bring your right foot back, and strike his head from above.

If he strikes your head thusly from above, then bring your left foot back, set him aside with your front point, shoot in strongly with your halberd above his, step to the inside with your right foot, and wind your rear point to his throat. If he sets you aside, then quickly pass forward with your left foot and strike your blade into his head.

If he attacks you with this same technique, then displace his strike between both hands onto your right side and strike him twice as you withdraw.

Halberd 11

Cod. Vinob. 10825, 183r

Inferus pectoris ictus contra ictum superum addito pariter vellendis habitu.

In congressum bipennis regendae hoc modo adaptabis te, dextrum pedem praepones, sursum inferum ictum pectoris reges versum eius brachium / sinistrum, atque in bipenne manus permuta, interim vero te de latere tuo dextro in latus sinistrum si converteris, versum te adtrahito.

In congressum bipennis regendae hoc modo adaptabis te, dextrum pedem praepones, sursum inferum ictum pectoris reges versum eius brachium / sinistrum, atque in bipenne manus permuta, interim vero te de latere tuo dextro in latus sinistrum si converteris, versum te adtrahito.

Sin autem adversarius idem contra te conetur, pedem sinistrum praefigentem, manum sinistram subito in bipenne tua transmutabis, et caput / eius ferias superne. Cum vero hostis eum ictum exceperit mucrone anterioris, tum sursum inflectes contra pectus eius in latus eius / sinistrum.

Si tu impulsum animadverteris ab adversario factum, pede dextro reducto, mucrone posterioris eum repelles. Interea / etiam laborabis de alio latere in aliud mucrone eodem, dextro insequitor, inter brachium eius utrumque bipennem tuum inferire contra pectus / hostis memineris, et eo habitu bipennem tuam supra brachium eius sinistrum converte. At si hostis manu a sua bipenne non removerit, / brachium eius infringes.

Cum vero tibi brachium eadem ratione confringere fuerit molitus, manum tuam mutabis, atque pede dextro / intro insequutus, nec non mucrone posteriori hostis pudenda impulsu concusseris, ab eo sine negotio te liberabis.

C93, 207r

Ein under er prusthaw gögen ainem oberhaw mit ainem reyssen.

Item schickn dich also in das zufechten stand mit deinem / rechten fuoß vor unnd gee auf mit dem unndern prusthaw im zü seinem lincken arm unnd / verwechsel deine hend an deiner hellenparten in dem wind dich von deiner rechten aüf dein lincken / seiten unnd reiß damit an dich.

Reist er dich also mit der sterckn zü im unnd dü mit deinem lincken / fuoß vorsteest so wechsel dein lincken hand behend durch an deiner hellenparten und haw im oben / zü seinem haüpt verseßt er dir das mit seinem vordern orth so wind im von unnden zü seiner prust aüf / sein reicht seiten.

Wirstü des stoß gewar so seß deinen rechten fuoß zü rückn unnd nimb das ab mit deinem / hindern ort indes arbait mit dem selbigen ort von ainer seiten zü der anndern volg mit deinem rechten / schenckel hinnach unnd wind im dein hellenparten zwischen seinen baiden armen hinnein zü / seiner prust unnd wend damit dein hellenparten ober seinen lincken arm last er dann sein hand / nit von seiner hellenparten so prichstü im den arm.

Begert er dir den arm also zü prechen so verwechsel / dein hand an deiner hellenparten indes trit mit deinem rechten fuoß hinein unnd stoß / im mit deinem hindern ort zü seinem gemechten so magstü damit sein ledig werden.

Halberd

A low *prusthaw*[1] with a pull against an *oberhaw*.[2]

Position yourself as follows during the approach: place your right foot forward, strike a low *prusthaw* up into the enemy's left arm, switch your hands on the halberd, turn your body from your right side to your left, and pull him toward you (as depicted by the left figure).

If he pulls you thusly toward him while you are standing with the left foot forward, then quickly change your left hand on your halberd and strike his head from above (as depicted by the right figure). If he parries your strike with his forward point, then wind up into his chest from underneath on his left side.[3]

If you anticipate a thrust from the adversary, then bring your right foot back, set him aside with your rear point, work with this same point from one side to the other, pass forward with your right foot, strike your halberd between both of his arms into his chest, and then wind it over his left arm; if he does not release his hand from his halberd, then you will break his arm.

If he attempts to break your arm with this same technique, then move your hand, step to the inside with your right foot, and thrust your rear point into his groin to free yourself from him.

---

1. German, "breast strike." See Latin *ictus pectoris*.
2. German, "over strike; strike from the upper openings." See Latin *ictus superus*, "above strike."
3. The German reads *reicht seiten*, or "right side."

Halberd 12

Cod. Vinob. 10825, 183v

Duo habitus vellendi superne ex primo bipennium contactu, vel collisione.

Quum ad invicem congressi bipenne athletice gubernanda fueritis, bipennibusque invicem contactis, pedem dextrum praepones, atque laminam / adversariis collo adiunges anterius.

Sin autem is ea ratione te adgressus fuerit lamina suae bipennis, sinistrum pedem praefigentem, vicissim / tuam bipennem collo eius adplicabis, atque in ipsa inflexione firmiter pro viribus versum te adtrahito.

Verum si hostis habitu eo fuerit usus / pede sinistro intro contra hostem concedas, atque eius bipennem excutito mucrone posteriori in tuum latus sinistrum, subito autem mucronem / eundem colli parti anteriori adiungito, dextrum pedem eius sinistro postponas, et si versus dextrum tuum latus convulseris humi prosternes / adversarium.

Sin autem idem contra te perficere statuerit, ab bipenne tua manum sinistram removeto, et si eadem partem brachiis supra / cubitum dextrum adpraehenderis, hostemque repuleris, liberum ea habitu te constitues. Confestim vero bipennem rursum arripe sinistro / caput eius feries, atque mutando ab hoste recedas.

C93, 207v

Zway obere reyssen auß dem anpund.

Item wann ir baid mit dem zufechten zu sainen kument / unnd ain ander ange binden hapt, so trit mit deinem rechten fuoß fur deinen lincken unnd wind im / dein plat vornen umb seinen hals.

Begert er dich also zu reissen mit seinem plat unnd du mit deinem / lincken fuoß vorsteest so wind im auch mit deiner hellenparten umb seinen hals unnd inn dem winden / so raiss starkn an dich.

Reist er dann also starckn an sich so trit mit deinem lincken schenckel hinnein / unnd nimb sein hellenparten mit deinem hindern ort hinweckn auf dein linke seiten inn dem far / im mit deinem hindern ort vornen umb seinen hals unnd seß deinen rechten schenckel hinder seinen / lincken reiss damit auf dein recht seiten so magstu in werffen.

Begert er dich also zu werffen so lass / dein lincke hand von deiner hellenparten unnd greiff damit wol hinder sienen recht elenpogen / scheub damit von dir so bistu sein ledig indes greiff behend mit deiner lincken hand wider nach deiner / hellenparten unnd schlag im nach seinem haupt in dem wechsel dich damit von im zu ruckn.

Two upper pulls from the bind.

When you and the enemy approach each other and bind halberds, place your right foot forward and hook your blade around his neck (as depicted by the left figure).

If he tries to pull you with his blade in the same manner while you are leading with the left foot, then hook your blade around his neck and pull it toward you with all your might (as depicted by the right figure).

If he employs this same technique, then step to the inside with your left foot, push his halberd off to your left side with the back point, bring that same point to his neck, place your right foot behind his left foot, pull toward your right side, and throw him to the ground.

If he tries to throw you, then release the halberd with your left hand and seize his right arm above the elbow, push him off, bring your left hand back to your halberd, strike his head, and withdraw from him with a change.

Halberd 13

Cod. Vinob. 10825, 184r

Habitus vellendi inferne contra impulsum violentum.

Hic modo te componas, sinistrum pedem praepone, iuxta pedem sinistrum bipennem contineas, laminam contra adversarium. Sed si hostis / contra te in habitu impulsus violenti constiterit laevum pedem praefigens, et conatur te figere, tum circum pedem eius, tuam bipennem anterius / applica, ea modo si de latere tuo sinistro dextrum versus deflexeris, hostem velles, ita ut eius impulsus tibi obesse nequeat, simulque prosterne / eum poteris.

Sin autem is eadem fuerit contra te molitus, tum repente tuam bipennem ab eius collo remove, verum tuae bipennis / laminam, eius bipennis adplica, firmiterque si suppresseris, te liberabis, ne ab hoste prosterni queas. Interim vero ab eius bipenne tuam / sursum tollas contra eius visum. Caeterum si hostis id removerit se defendendo, bipenne rotata de latere eius sinistro in tuum / dextrum, collam eius impulsu quassabis. Sed si id rursum repulerit, pede dextro introgreditor, atque mucronem posteriorem versus / adversariis visum convertas.

Si autem is eadem usurpet, sinistrum pedem referre non dubites, et eius mucronem anteriorem posteriori / tuae bipennis repellas, subito autem sinistro pede intro concedas, mucronem anteriorem versus hostis visum si direxeris, in habitu / mutatoriis ab adversario recedas.

C93, 208r

Ein unders reyssen gögen ainem gewalt stoß.

Item schickn dich also Stannd mit deinem lincken fuoß vor / unnd hallt dien hellenparten wol bey deinem lincken schenckel das plat gögen dem man stat er dann also / gögen dir Inn dem gewalt stoß seinen lincken fuoß furgeseßt unnd begert auf dich zü stossen so gee mit deiner / hellenparten Im vornen umb seinen schenckel unnd wend dich damit von deiner lincken auf dein rechten / seiten so reisstu In das dir sein stoß kainen schaden bringen kan und magstu In auch werffen.

Hat er / dich also gefast unnd begert dich zu werffen so laß dein hellenparten behend von seinem hals unnd seß Im / Indes mit dem plat auf sein hellenparten unnd truckn starckn unnder sich so bistu des wurff ledig In dem / far Im an seiner hellenparten behend auf zü seinem gesicht verseßt Er dir das so wechsel Im durch / von seiner lincken auf dein recht seiten unnd stoß Im nach senem hals verseßt er dir das furbas so trit / mit deinem rechten fuoß hinnein unnd wind Im deinen hindern ort zü seinem gesicht

Arbait er dir also / zü so seß deinen lincken fuoß zü rückn und nimb Im seinen vordern ort mit deinem hindern hinweckn / Indes trit mit deinem lincken fuoß hinnein und wind Im deinen vordern ort zü seinem gesicht zeuch / dich damit Inn dem wechsel zü rückn.

A low pull against a violent thrust to the face.

Position yourself in this manner: stand with your left foot forward and hold the halberd near your left leg, with your blade facing the adversary. If he then positions himself for a violent thrust to your face while leading with his left foot (as depicted by the right figure), then hook your halberd around his left leg, turn yourself from your right to your left side (as depicted by the left figure), pull him in such a way that his thrust cannot hurt you, and then throw him.

If he hooks you thusly and tries to throw you, then quickly remove your halberd from his throat, hook your blade around his weapon, free yourself by pushing down strongly, and slide your halberd up his shaft into his face. If he displaces you, then change him through from his left side to the right and thrust into his throat. If he parries that, too, then step to the inside with your right foot and wind your back point into his face.

If he works toward you thusly, then bring your left foot back, parry his front point with your back point while passing to the inside with your left foot, wind your front point into his eyes, and withdraw from him in *wechsel*.[1]

---

1. German, "changer." See Latin *habitus mutatorius*, "change guard."

Halberd 14

Cod. Vinob. 10825, 184v

Intorsio simul habitu addito adversariis prosternendi.

Quum congressi sitis bipenne athletice regenda, pariter bipennes invicem eodem momento contigeritis, ab eius bipenne tuam circumvolues / de latere adversariis dextro sinistrum versus, et impulsu brachium eius sinistrum pulsato. Verum si hostis id animadverterit, / et removerit tuum conatum, dupliciter bipennem circumvolues, atque proximas acquirito eius nuditates.

Sin vero is tuas eadem / ratione observarit, tum in ipsius conspectu tuam bipennem sursum vibrabis, Sed in ipso vibrandi habitu, brachium hostis anterius concuties, / et firmiter eius bipennem abtingito, si igitur quam celerrime eam circumvolueris, gemino impulsu visum eius adpetito.

Ceterum / si te ea ratione adgreditur, mucrone anteriori tuae bipennis repellito, sinistrum praeponens. Sin autem vicissim contra adversarium / constiteris laevum itidem praefigens, tum eius bipennem mucrone posterioris removeto, dextrum pedem hostis laevo postponas, atque mucronem / posteriorem collo eius adplica anterius, et si fortiter superne versus te adtraxeris, adversarium per dextrum pedem / tuum prosternes.

C93, 208v

Ain einwinden mit ainem wurff.

Item wann Ir baid mit dem zufechten zu samen / kumment unnd gleich mit ain annder angepunden hapt so wechsel Im durch an seiner hellenparten von / seiner rechten auff sein lincken seiten unnd stoß Im damit nach seinem lincken arm wirt er des stoß / gewar unnd verseßt dir den so wechsel Im zwifach durch unnd süch Im seine nechste ploß.

Sücht er dir deine / plossen also so streich mit deiner hellenparten von unnden aüf Im vor seinem gesicht und In dem aüf / streichen haw Im nach seinem vordern arm Indes pind Im starckn an sein hellenparten unnd wechsel / Im behend durch unnd stoß Im zwifach zü seinem gesicht.

Stost er dir also zwifach zü so nimb Im das ab mit / deinem lincken fuoß vor so nimb sein hellenparten mit deinem hindern ort hinweckn trit mit deinem / rechten fuoß hinder seinen lincken unnd seß Im deinen hinden ort vornen umb seinen hals zeüch damit / oben starckn an dich so wirfstü In ober deinen rechten schenckel.

Halberd

An inside wind with a throw.

When you both meet in a simultaneous bind, change your opponent through from his right side to his left, and thrust into his left arm. If he anticipates your thrust and parries, then change him through twice and find his nearest opening.

If he probes for your openings thusly, then extend your halberd toward his face; during this extension, strike his leading arm, bind his weapon strongly, quickly change him through, and thrust twice into his eyes.

If he thrusts at you twice in the same manner while your left foot is leading, then parry him with your front point, push his halberd away with your back point, place your right foot behind his left foot, hook your back point around his neck (as depicted by the right figure), pull his upper body strongly toward you, and throw him over your right leg.

Note: This illustration, along with the physical mechanics of the technique, suggests that the position of the halberds in Shortstaff 10 is an artistic error.

107

Halberd 15

Cod. Vinob. 10825, 185r

Supera transitio adhibito lateris impulsu, ut infra describetur.

In eum scilicet modum iam praenominatum bipenne rite gubernanda, te ea ratione, qua mox describetur, adaptabis, sinistrum pedem praepones / in latere dextro bipennem tuam continebis, mucro sit contra adversarium porrectus. Et si is vicissim contra te in habitu corporis eodem / constituerit, simul eam adversario ferias, et in bipennium contactu, laminam tuam retro adversariis bipennis mucronem recurvum adplicabis, / sursumque ex ea forma attolles.

Sin autem bipennem tuam eadem ratione comprehenderit, eodemque modo usus fuerit, tum dextrum pedem hostis / sinistro postpones. Verum subito tuam bipennem superne retrahe, ac inferni mucronem posteriorem eius lateri sinistro impellito. Caeterum / dextro pede reducto laminam tuae bipennis adversariis capiti infligito.

Si vero is eadem contra te fuerit molitus, mucrone posteriori eius / conatum repelles, interea autem dextro insequutus, brachium sinistrum hostis sauciabis. Porro si id exceperit adversarius, tum inferas / mucronem posteriorem intra ipsius brachium utrumque versus pectus, inde vero dextro pede retrorsum concedas, et si caput adversariis / quassaris, ab adversario feriundo recedes.

C93, 209r

Ain obers ubergeen mit sampt ainem seiten stoß.

Item schictn dich Allso mit disem zufechten stannd mit / deinem lincken fuoß vor unnd hallt dein hellenparten aüf deiner rechten seiten den Ort gögen / dem man stat er dann auch also gögen dir In dem zufechten so pind mit Im gleich an unnd in dem / anpund wind dein plat hinder seinen haggen unnd scheuß damit uber sich.

Hat er dir dein hellenparten / also gefast unnd scheubt dich ober sich so trit mit deinem rechten fuoß hinder senen lincken / Inn dem zuckn dein hellenparten oben unnd stoß Im unnden mit deinem hindern ort zü seiner / lincken seiten indes seß deinen rechten fuoß wider zü rückn unnd haw Im mit deinem plat zü seinem / haupt.

Arbait Er dir also zü so nimb Im das ab mit deinem hindern ort Indes volg mit / deinem rechte fuoß hinnach unnd haw Im nach seinem lincken arm verseßt Er dir das so / wind Im mit deinem hindern ort zwischen baiden arm hinnein zü seiner prüst Indes trit mit / deinem rechten schenckel zü ruckn unnd haw Im nach seinem haüpt so hawstu dich von Im.

An *obergehen*[1] with a thrust to the side.

Position yourself in the approach as follows: place your left foot forward and hold the halberd on your right side, with your point extended toward the opponent. If he in turn stands against you in the same position, then immediately bind his halberd, wind your blade around his hook, and lift upward.

If he hooks your halberd and lifts it up thusly, then place your right foot behind his left foot, immediately pull your halberd back and upward, and thrust your rear point into his left flank from underneath (as depicted by the right figure); furthermore, bring your right foot back and strike your blade into his head.

If he tries to do the same thing to you, then parry his thrust with your back point, pass forward with the right foot, and strike his left arm. If he parries, then wind your rear point between his arms into his chest, bring your right foot back, and strike his head as you withdraw.

---

1. German, "going over." See Latin *transitio*, "passing over."

Halberd 16

Cod. Vinob. 10825, 185v

Ictus inferi habitus cum aversione per bipennem medium.

    Contra adversarium bipenne gubernanda athletice hoc pacto te accomodabis, sinitrum pedem praeponito, bipennem tuam in dextro latere contineas. / Sin vero adversarius contra te eodem modo constiterit in habitu ictus inferi dextrum praeponens, et bipennem suam in latere sinistro / continens, tum versus faciem hostis bipennem vibrabis.
    Si autem is eam vibrandi rationem contra te usurparit, eius sinistrum brachium sursum / feriundo quassato.
    Porro cum is brachium sinistrum tibi adpeti[v]erit, tum bipenne media removebis ipsius conatum versus latus tuum dextrum, / et ex eo habitu superne lamina tuae bipennis eius brachio sinistro adplicata, supprimito.
    At si is adeo firmiter suppresserit, sinistro pede / retro hostis dextrum collocato, mucronem posteriorem capiti eius infligito, sed celeriter pedem dextrum si reduxeris, contra pectus adversariis / mucronem anteriorem propellito.
    Caeterum si is idem usurparit, tum media bipenne eum repellas, atque mucronem posteriorem / inferne supra brachium sinistrum hostis inflectes, et si in latus tuum dextrum convulseris, in habitu ictus supernis ab adversario recedes.

C93, 209v

Ein underhaw mit ainem abseßen In der halben hellenparten.

    Item schickn dich also Gegen dem man In dem zufechten / stand mit deinem lincken fuoß vor unnd hallt dein hellenparte aüf deiner rechten seiten stat er dann / aüch also gögen dir Inn dem unnderhaw seinen rechten fuoß furgeseßt die hellenparten an seiner lincken / seiten so streich mit deiner hellenparten Im zü seinem gesicht.
    Geet er dir also zü deinem gesicht so haw Im / von unnden nach seinem lincken arm.
    Hawt er dir also zü deiner lincken seiten so seß Im das ab mit deiner / halben hellenparten aüf dein rechten seiten unnd In dem seß Im dein plat oben aüf seinen lincken arm / unnd truckn damit unndersich.
    Truckn Er dann also undersich mit sterckn so trit mit deinem lincken / schenckel hinder seinen rehcten unnd haw Im mit deinem hindern orth zü seiner prüst.
    Windt Er dir also / ein so seß Im das ab mit deiner halben hellenparten unnd wind Im deinem hindern ort under ober / seinen lincken arm reiß damit aüf dein rechten seiten unnd zeüch dich mit ainem oberhaw von Im / zü rückn.

Halberd

An *underhaw*[1] with a displacement using the middle of the halberd.

Position yourself against the opponent in this way: place your left foot forward and hold your halberd on the right side (as depicted by the left figure). If he in turn takes position against you for an *underhaw* from the left side with his right foot leading, then thrust your halberd into his face.

If he thrusts thusly at your face, then strike an *underhaw* into his left arm (as depicted by the right figure).

If he strikes your left arm in this same way, then deflect him onto your right side with the middle of the halberd, set your blade onto his left arm, and press down.

If he presses your arm down thusly, then place your left leg behind his right, strike your back point into his head, bring your right foot back, and thrust your front point into his chest.[2]

If he tries to do the same thing to you, then set him aside with the middle of your halberd, wind your back point up over his left arm, pull toward your right side, and withdraw from him with an *oberhaw*.

---

1. German, "under strike; strike from the lower openings." See Latin *ictus infernus*, "lower strike."
2. The German simply reads "step with your left leg behind his right and strike him with your back point to his breast."

Halberd 17

Cod. Vinob. 10825, 186r

Incisio, addita forma hostis prosternendi ex habitu pedis inisciendis in adversariis poplitem.

In formam, quae infra mox describetur hoc modo te adaptabis ex superioris incisione, dextrum pedem praeponito, supra caput bipennem / contineto, lamina eiusdem sursum flexa.

Et si contra adversarium dextro pede praefixo constiteris, atque in latere dextro bipennem contineris / versus hostis visum, sinistro intro procedas, laminam collo eius anterius applica.

Sin autem hostis modo eodem contra te fuerit / usus, tum mucronem tuae bipennis posteriorem sub alam hostis sinistram ingeras, versum latus tuum dextrum propellito, verum dextrum / pedem si retraxeris, lamina caput hostis proscindes superne.

Sed si te conabitur proscindere, bipenne sursum elevata supra caput, eius / incisionem excipias tuae bipennis lamina, inde autem dextrum pedem hostis pedi sinistro postpones, atque eius bipennem superne mucrone / posteriori reiscito.

Cum vero is te eadem ratione repulerit, sinistro pede reducto mucronem posteriorem versus brachium adversariis / sinistrum inflectito.

Hoste idem usurpante, dextro eius pedi sinistrum postpones, atque laminam collo hostis anterius applicabis, / pedem eius dextrum stimulo recurvo bipennis quassato exterius, interim vero ex eo habitu fortiter superne eum protrudes, inferne / eum si versum te adtraxeris, prosternis adversarius cogetur.

C93, 210r

Ain schnit mit ainem Wurff auß dem hacknen.

Item halt dich also Mit dem zufechten auß dem obern / schnit stand mit deinem rechten fuoß vor unnd halt dein hellenparten ob denem haupt das plat / ober sich.

Gewendt stastu dann auch also gögen Im deinen rechten fuoß furgeseßt so halt dein hellenparten / auf deiner rechten seiten im zu seinem gesicht n dem trit mit deinem lincken schenckel hinein / unnd seß im dein plat vornen seinen hals.

Hat Er dir also angeseßt so seß im deinen hindern / ort under sein lincke achsen scheub damit auf dein rechte seiten Indes trit mit deinem rechte fuoß / zü ruckn unnd schneid Im mit deinem plat oben nach senem haupt.

Schneidt er dir also nach deinem / haupt so far auf mit deiner hellenparten wol fur dein haupt unnd verseß Im den schnit mit deinem / plat In dem trit mit deinem rechten fuoß hinder seinen lincken unnd nimb Im sein hellenparten / oben mit deinem hindern Ort hinwechtn.

Nimpt er dir dein hellenparten allso hinwechtn / so trit mit deinem lincken fuoß zü ruckn unnd wind Im mit deinem hindern Ort nach seinem / lincken arm.

Windt Er dir also zü so trit mit deinem lincken fuoß hinder seinen rechten und seß / Im da plat vornen an seinen hals und schlag Im den haggen auß wendig umb seinen rechten / fuoß Indes scheub oben starckn von dir und zeüch unden zü dir so muoß er fallen.

A cut with a throw from a kick to the heel.

Position yourself thusly from a high cut during the approach: place your right foot forward and hold the halberd above your head, with its blade turned upward.

If you position yourself against the opponent with your right foot forward and the halberd on your right side, pointing at his face, then step to the inside with your left foot and press your blade to his throat (as depicted by the right figure).

If he does the same thing to you, then force your back point under his left armpit (as depicted by the left figure), push toward your right side, bring your right foot back, and cut his head open with your blade from above.

If he tries to cut you in this way, then raise your halberd up above your head, displace his cut with your blade, place your right foot behind his left foot, and push his halberd away with your rear point.

If he pushes your halberd away thusly, then bring your left foot back and wind your back point into his left arm.

If he employs this same wind against you, then place your left foot behind his right foot, press the front of your blade to his throat, strike your halberd's curved spike to the outside of his right foot, strongly push him over away from you while pulling him toward you from underneath, and throw him.

Halberd 18

Cod. Vinob. 10825, 186v

Habitus quo adversarius convertitur contra ictum inferum.

Ad hunc modum ita te adaptabis, convertes te pede dextro in latus tuum sinistrum in habitu ictus superni ita, ut pes tuus laevus / contra adversarium sit conversus, atque ex ea forma caput eius ictu superno concutias.

Verum si hostis idem usurpaverit contra / te sinistrum pedem praeponentem, et mucro tuae bipennis posterior terrae sit inclinatus contra hostem, tum eandem sursum levabis, / et eius ictum eodem repellas mucrone. Porro dextro pede intro progressus, feriundo ex ictu infero brachium adversariis sinistrum / conquassabis.

Sin autem idem exceperit, mucronem bipennis tuae posteriorem in latus tuu dextrum dirigas, et si pede dextro / retrorsum concesseris, mucronem anteriorem de latere eius sinistro versus dextrum rotabis contra hostis visum.

Eo autem habitu / consimili utente, parte anterioris bipennis, eius conatum repelles, verum ab eius bipenne si tuam circumvolueris de latere alio in / aliud, nuditates eius pervestigato. Si vero adversarius id removerit, dextro pede intro concedas, atque eius bipennem tuae / bipennis mucrone posteriori reiscias, et si eam vultus eius vel pectori impegeris in habitu mutatoriis ab hoste recedas gradu gemino.

C93, 210v

Ain verkerer gögen ainnem underhaw.

Item schickn dich also In den verterer verwend dich auf / deinen rechten fuoß auf dein lincke seiten herumb In dem oberhaw das dein lincken fuoß / gögen dem man gewendt sey In dem haw Im auß dem oberhaw zu seinem haupt.

Hawt Er / dir also oben zu unnd du mit deinem lincken fuoß vorsteest deinen hindern orth vornen / auf der Erden gögen dem man so far auf mit dem seibigen ort unnd nimb Im damit / den haw hinweckn In dem trit mit deinem rechten schenckel hinein unnd haw Im auß / dem underhaw von unden nach seinem lincken arm.

Verseßt er dir das so gee mit deinem / hindern ort auf dein rechten seiten indes trit mit deinem rechten schenckel zu ruckn unnd / wechsel Im deinen vordern orth von seiner lincken auf sein rechten seiten zu seinem gesicht.

Wechselt er dir also zu deinem gesicht so verseß im das vornen an deiner hellenparten in / dem wind Im durch an seiner hellenparten von ainer seiten zu der andern und such Im / die nechste ploß seßt er dir das ab so trit mit deinem rechten schenckel hinein und nimb / Im sein hellneparten mit deinem hindern ort hinwechtn stoß Im damit nach seinem / gesicht oder der prust Indes trit In dem wechsel zwifach zu ruckn.

Halberd

A technique in which the adversary is turned around, against an *underhaw*.[1]

Position yourself thusly for this technique: hold your halberd in the guard for an *oberhaw*,[2] turn yourself around with your right foot on your left side in such a way that your left foot is turned toward the adversary (as depicted by the left figure), and strike an *oberhaw* into his head.

If he strikes you thusly from above when you are standing with your left foot forward and your back point tilted toward the ground (as depicted by the right figure), then raise your halberd up high, parry his strike with your back point while stepping to the inside with your right foot, and strike an *underhaw* into his left arm.

If he parries you in the same way, then bring your back point onto your right side, step backward with your right foot, and change him through with your front point, from his left to his right side and into his eyes.

If he changes you through thusly, then displace him with the front part of your halberd, wind around him from one side to the other, and seek out his openings. If he parries, then step to the inside with your right foot, displace him with your back point, thrust it into his face or chest, and take two steps away from him in *wechsel*.[3]

---

1. German, "over strike; strike from the upper openings." See Latin *ictus superus*, "above strike."
2. German, "under strike; strike from the lower openings." See Latin *ictus infernus*, "lower strike."
3. German, "changer." See Latin *habitus mutatorius*, "change guard."

Halberd 19

Cod. Vinob. 10825, 187r

Ratio, qua adversario bipennis eripitur adhibita forma eiusdem prosternendi.

Si ad invicem fueritis congressi bipenne athleticem regenda, et pariter bipennes mutuo contigeritis, tum tuam ab hostis bipenne circumvolues / de latere eius dextro versus sinistrum impulsu adhibito. Si vero adversarius te repulerit, dextro pede intro si concesseris, / eius bipennis locum latiorem mucrone posteriori repelles versus latus tuum dextrum.

Caeterum si idem perficere is conabitur, sinistro pede / reducto, mucronem anteriorem pectori hostis impellito. Sed si exceperit adversarius, tum gradu duplici contra ipsum procedas, / et mucronem posteriorem eius capiti infligito.

Verum si idem is molitus fuerit contra te, media bipenne tua eum repellas. Interea / autem hostis bipennis mucronem posteriorem manu sinistra si adpraehenderis, dextra vero anteriorem, ita ut utramque bipennem tu / contineas euis manibus, pedem sinistrum, eius genu dextro si imposueris, confringere pedem eius licebit.

Si autem is bipennem utramque / adpraehenderit, praedictumque modo habitum usurparae volet, dextram manum de tua bipenne removebis, atque dirige exterius circum / eius poplitem, manu autem sinistra firmiter superne propellito hostem, inferne vero sursum si sustuleris, concidere eum coges, / atque eius conatus irritus omnis fiet.

C93, 211r

Ain hellenparten nemen mit ainem wurff.

Item wann ir mit dem Zufecthen zü samen kument und gleich / mit ain annder angepunden hapt so wechsel Im durch von seiner rechten auf sein lincken seiten mit / ainem stoß Nimpt er dir das ab so trit mit deinem rechten schenckel hinnein unnd Nimb Im sein / plat mit deinem hindern ort hinweckn auf dein reckte seiten.

Begert er dir dein hellenparten also hinweckn / zü nemen so trit mit deinem lincken fuoß zü rückn unnd stoß Im deinen hindern ort zü seiner / prüst verseßt er dir das so trit widerumb zwifach zü Im hinein und schlag Im mit deinem hinder ort / zü seinem haupt.

Schlecht Er dir also nach deinem haupt so verseßt Im das miten in deiner hellenparten / Indes greiff mit deiner lincken hand nach seinem hindern Ort und mit der rechten nach / seinem vordern das du die baid hellenparten In deinen henden hab Ist In dem trit mit deinem linck / fuoß auf sein rechtes knie so magstu Im den schenckel entzway prechen.

Hat er dir die baid hellenparten / also zü samen gefast und begert dich zü treten so laß dein rechte hand von deiner hallenparten und / far Im auf wendig umb sein kniebug Indes scheub oben mnit deiner lincken hand starkn von dir und / zeuch unden fast ober sich so muß er fallen unnd kan dir sein arbait mit schaden.

A halberd disarm with a throw.

When you bind halberds with the opponent, change him through from his right to his left side, and thrust. If he parries you, then step to the inside with your right foot and push his blade onto your right side with your back point.

If he tries to do the same thing to you, then bring your left foot back and thrust your forward point into his chest. If he parries, then take two steps toward him and strike his head with your back point.

If he strikes your head thusly, then displace him with the middle of your halberd, simultaneously seize his back point with your left hand and his front point with your right hand so that you are holding both halberds at once, step onto his right knee with your left foot (as depicted by the left figure), and break his leg.

On the other hand, if he seizes both halberds and tries to break your leg, then let go of your halberd with the right hand, quickly reach it around his knee (as depicted by the right figure), lift up while strongly pushing his upper body with your left hand, and throw him.

Halberd 20

Cod. Vinob. 10825, 187v

Ratio, qua pes adversario infringi potest contra impulsum eum quo cervix impetitur.

Quum contigerit, ut ad invicem prope processeritis more athletico, et insimul bipennes mutuo contigeritis, bipennem tuam contra hostis visum / contineas, verum ab eius bipenne tua circumvoluta, visum eius impulsu adgreditor. Sin autem tuum impulsum animadverterit, averteritque, / dextro pede intro procedas, atque mucronem posteriorem contra eius brachium sinistrum ex loco infero convertito.

Si autem is idem fuerit / molitus contra te, in bipenne tua celeriter manum sinistram mutabis, pectusque eius mucrone anterioris quassato. Sed si id exceperit, / mucronem posteriorem capiti hostis infligito.

Caeterum si is superne te ratione adgressus fuerit eadem, retrorsum dextro pede recedas / atque ipsius ictum mucrone anteriori removeto, interim vero fortiter bipennis eius partem inferiorem contingas, eoque habitu si sustuleris, / duplici gradu contra eum intro procedas, bipennem abisce, sinistra eius pedem inferne manu dextra adpraehendas, manu vero / sinistra eius molam acriter urgeas, et eo modo pedem eius confringere licebit, vel hostem prosternere.

Porro si te coartarit, cervici / eius tuam bipennem adplicato, et si acriter suppresseris, eius conatum omnem irritabis.

C93, 211v

Ein bainpruch gegen ainem nactstoß.

Item wann ir mit dem zufechten zu samen kument / unnd gleich mit ain ander angepunden hapt so halt dein hellenparten im gegen seinem gesicht / in dem wechsel ab an seiner hellenparten unnd stoß im damit zu seinem gesicht wirt er des stoß / gewar unnd verseßt dir den so trit mit deinem rechten fuoß hinein unnd wind im deinen hindern / ort von unnden nach seinem lincken arm.

Windt er dir also unden zu so verwechsel dein / lincke hand behend an deiner hellenparten unnd stoß im mit deinem vordern ort nach seiner / prust verseßt er dir das so schlag im deinen hindern ort zu seinem haupt.

Schlecht er dir also / oben ein so trit mit deinem rechten fuoß zu ruckn unnd nimb seinen haw mit deinem vordern / ort hinweckn in dem pind im starkn under sein hellenparten unnd schub damit obersich indes / trit zwifach zu im hinnein unnd laß dein hellenaprten fallen greiff mit deiner rechten / hand unden an seinen lincken fuoß unnd mit deiner lincken gewaltig auf sein knie / scherben so magstu im den schenckel prechen oder werffen.

Hat er dich also gefast so seß im dein / hellenparten auf seinen nackn und truckn damit starckn undersich so prichtstu im sein arbeit.

A leg-break against a thrust to the neck.

When you bind halberds with the opponent, point your halberd toward his face, change him through, and thrust into his eyes. If he anticipates your thrust and sets it aside, then step to the inside with your right foot and wind your back point from underneath into his left arm.

If he winds you thusly, then quickly switch your left hand on the halberd and thrust your forward point into his chest. If he displaces you, then strike his head with your back point.

If he strikes your head from above in the same way, then bring your right foot back, parry him with your forward point, strongly bind the bottom of his halberd, lift it up, take two steps toward him on the inside, drop your halberd, and break his leg or throw him by reaching behind his left foot with your right hand while forcefully pushing his knee with your left hand (as depicted by the left figure).

If he closes with you thusly, then set your halberd on his neck (as depicted by the right figure) and press down strongly to neutralize his efforts.

Various 21

Cod. Vinob. 10826, 120r

Custodia cancellata contra infernam.

Ad hunc habitum ea ratione te accommodato, versus hostem sinistrum pedem praeponas, atque ex inferna custodia sursum punctione / directa visum adversariis fodias.

Quod si is idem usurparit, te in custodia cancellata consistente, dextrumque praeponentem / pedem contra hostem, tum sinistrum intro colloces. Verum in ipso passu, ictus ipsius obviam ferias de superne, post si suppresseris, / punctionem hostilem avertes.

Sin autem idem is molitur, tum lamina tuae bipennis, hostis bipennem retorqueas, / de latere tuo sinistro, versus adversariis latus itidem sinistrum, et ea ratione intercludes hostem, inde arrepta bipenne, visum / ipsius vel pectus adpetas.

Sed si is pari ratione te adgreditur fodiendo, tum sinistrorsum impetum hostilem, lamina tuae bipennis / avertas, postea laminam capiti eius infligas, atque gemino passu ab ipso recedas.

Cod. Vinob. 10826, 120r

Die geschrenckt hüt gegen der undern.

Item halt dich also mit disem stück in dem zufeckten, Trit mit deinem linggen schenckel zü Im / hinein, und gee aüs der undern hüt aüff mit ainem stich Inn zü seinem gesicht.

Sticht er / also aüf dich von unden, und dü in der geshrenckten hüt gegen Im steest, mit deinem rechte füöß / vor, so trit mit deinem linggen schenckel hinein, und in trit häw von oben seinem stich entgegen, / truckn damit wol undersich, so nimbst dü Im seinen stich hinweckn.

Hawt er dann also aüf / dich, und hat die deinen stich abgenomen, so wind mit deinem plat an seiner hellenbarte / von deiner linggen aüf sein linggen seiten, so spertz dü Im In dem so zück den hellenbarre / und stoß Im nach seinem gesicht odere der prüst.

Sticht er also aüf dich, so nimb Im das ab / mit deinem plat aüff dein linggen seiten, Indes häw Im mit deinem plat zü seinem / haübt, wind dich damit zwifach von Im zu ruckn.

Halberd

The crossed guard against the low guard.

For this technique, position yourself as follows: stand in a low guard with your left foot forward, toward the enemy (as depicted by the left figure), and thrust straight up into his face.

If he thrusts at your face from the low guard while you are standing in a crossed guard, your right foot forward (as depicted by the right figure), then step to the inside with your left foot, strike against his thrust from above, and displace him by pressing his weapon down.

If he binds your thrust thusly, then wind from your left side to his left side with your blade on his halberd, so that you block him; then, pull your halberd back and thrust into his face or chest.

If he thrusts at you in this same way, then set his thrust onto your left side with your blade, strike your blade into his head, and step twice from him into the withdrawal.

Note: This plate only appears in the "Various Weapons" section of the Vienna and Munich MSS.

Chapter 5

Poleax

"**SECURES** letales ab amazonibus item repertas esse ferunt, atque iis / usas, verum mirabile est auditu ab muliebri sexu tam letalia duo arma esse inventa, / multi etiam incredibile videret, ni Aventinus veritatis studiossimus, pluresque / aliis ea de re scripsissent, ea continet, 16, habitus" (Cod. Icon. 393, 15r, 6–9).

"**POLEAXES** are also said to have been invented by the Amazons, and used by them; in fact, it is astonishing to hear [that] two such deadly weapons were invented by the feminine sex; it may also seem incredible to many, except the most learned man of truth, Aventinus,[1] and many more have written in response concerning these same things. [This chapter] contains 16 techniques."

The poleax depicted in the *Opus* is of an especially vicious variety, hence Mair's use of the terms *secures letales*, or "deadly ax," and *mordagst*, or "murder ax," to describe it. The modular head consists of a spear-like front point, a diamond-shaped ax blade, a dagger-like blade on the reverse, and long spikes on both sides, perpendicular to the blades. There is also a spike at the butt end of the wooden haft. The weapon measures roughly four feet in length.

As with the halberd, Mair credits the development of the poleax to the Amazons, but it is actually a 14th-century European adaptation of the infantryman's battle-ax. It found widespread use among the knightly classes well into the 16th century, particularly in single combat on foot, as it was very effective against armor. Poleaxes varied widely in their design; the most common featured a hammer head opposite a flat or curved ax blade. The length of the weapon typically ranged from 4 to 6 feet.

Most, if not all, of the techniques from the preceding chapters are applicable here. However, with its shorter length, angular blades, and side spikes, the poleax is especially well suited for striking and thus shares many similarities with the longsword that staff and halberd do not, most notably in the footwork. Therefore, we strongly recommend that the reader become well versed in longsword fundamentals before advancing.

It should be noted that there is some debate over the etymology and correct spelling of the word "poleax," with some preferring the Middle English "pollax" on the grounds that the weapon was designed to crush "polls," or heads. We have chosen the modern spelling, which immediately distinguishes the weapon as a polearm.

---

1. Johannes Aventinus (1477–1534) was a Bavarian historian and philologist.

POLEAX 1

Cod. Vinob. 10826, 149r

Aversiones effigies contra plagam letalem per securim.

　Dextrum pedem in congressu contra hostem praeponito, supra caput tuum, securim utrisque manibus teneto, atque mucronem porrectim adversus / hostem. At si contra te is sinistrum pedem praefixerit, et caput tuum plaga letali concutere conetur, tum securi tua eam removeas in latus sinistrum, / inde autem retracta securi, pubem adversariis figas.

　Eodem vero habitu hoste utente, posteriori securis mucrone eam vim repellas. Post / autem insequutus pede dextro, lamina securis tuae faciem ipsius saucies. Verum si adversarius eum impetum hastili securis removeat intra / manum utramque, mucronem posteriorem intorqueas intra brachium utrumque contra hostilem faciem.

　Eodem modo si is te adgreditor, dextrum / pedem referto, atque hastili intra utramque manu hostem excipito. Sed si eveniat, ut tuam vim excipiat, tu repente supernam nuditatem / adpetas feriundo.

　Cum vero hostis te feriundo adgredi molitur, capulo tuae securis eum repellas, inde si in latus dextrum tuum / convulseris, faciem eius pungas mucrone, inde autem securi circumrotata recedas.

C94, 164r

Ein Abnemen In der mordagst gögen ainem morschlag.

　Item schickn dich also in Dises stuckn wann In zü samen gond / stannd mit deinem rechten fuoß vor unnd halt dein mordagst Inn baiden henden ob dein haupt In / der hoch den ort gestrackt gögen dem man stat er dann mit seinem lincken fuoß gögen dir und schlecht / dir ainem mordschlag nach deinem haupt so nimb Im das ab mit deiner mordagst aüf dein lincke / seiten Indes zuckn dein mordagst an dich unnd such Im zü seinem gemechten.

　Sucht er dir also unden zü / so nimb Im das ab mit deinem hindern Ort deiner mordagst Inn dem volg mit deinem rechten / schenckel hinnach und haw Im mit deinem plat deiner mordagst nach seinem gesicht verseßt er dir / das zwischen baiden henden den In sein gehulß so wind Im deinem hindern Ort ywischen seinem baiden / armen hinein zü seinem gesicht.

　Hawt unnd windt er dir also ein so trit mit deinem rechten schenckel zü / ruckn unnd nimb Im das ab zwischen baiden henden In dem gehulß verseß er dir das also so haw Im behend / widerumb nach seiner obern ploß.

　Hawt er dir also oben zü so nimb im das ab mit deinem gehulß / reiß damit auf dein lincke seiten unnd such Im zü seinem gesicht Inn dem trit und wind dich damit von / Im zü ruckn.

An *abnemen*[1] against a murder-stroke using poleaxes.

During the approach, place your right foot forward and hold the poleax above your head with both hands, your point extended toward the opponent (as depicted by the left figure). If he then stands against you with his left foot forward and attempts to strike your head with a murder stroke (as depicted by the right figure), then displace him with your poleax onto your left side, draw your weapon back, and thrust into his groin.

If he attacks your groin with the same technique, then take his attack aside with your back point, pass forward with your right foot, and strike your axe blade into his face. If he parries your attack between his hands on his haft, then wind your back point between his arms and into his face.

If he strikes and winds you thusly, then bring your right foot back and take him aside between your hands on your haft. If he displaces you, then quickly strike to his upper openings.

If he strikes you from above, then parry him with your hilt, pull onto your right side, thrust your point into his face, and withdraw from him while winding your poleax.

---

1. German, "taking-off." See Latin *habitus aversionis*, "parry; setting-aside."

Poleax 2

Cod. Vinob. 10826, 149v

Plaga letalis contra punctionem qua pectus pungitur.

Hac ratione in praescriptam pugnam te accommodes oportet, tibiis paribus erectus consistas, capulum securis tuae manu utraque contineas, sinistra / sit sublata retro caput tuum, at inde dextro pede si contra hostem processeris, de superne caput ipsius, securis lamina pulsato.

Sed si is pari contra / te modo fuerit usus, sinistrum praefigentem, secundumque caput securim tuam continentem dextra, sinistram vero adplicaris mucronis securis inferiori / adversus hostem, tum lamina securis tuae, adversariis ictum repelles in latus dextrum tuum, inde autem mucrone eius anterior pectus hostile / punges.

Verum si is idem efficere moliatur, hastili intra manum utramque eam vim excipias versus latus sinistrum, post mucronem anteriorem / securis tuae supra brachium sinistrum directum in pectus adversariis transmittas, atque fortiter adtrahas vellendo, et eo negocio confecto / regredi licebit securi rotata.

C94, 164v

Ain mordschlag gögen ainem prust stich.

Item wann Fr zü samen Gond so schickn dich also / in diseß stuckn stannd mit gleichen fuossen aüfrechte unnd halt dein mordagst mit baiden henden / Inn dem gehulß dein lincke hannd In der hoch hinder deinem haüpt In dem trit mit deinem rechten / schennkel hinein unnd haw Im mit deinem plat von oben nach seinem haüpt.

Hawt er also von / oben aüf dich unnd dü mit deinem lincken füoß gögen Im steest und helst dem mordagst mit deiner / rechten hand hinder deinem haüpt dein lincke bez deinem vordern Ort gögen dem man so nimb / Im seinem haw mit dem plat deiner Mordagst hinweckn aüf dein recht seiten Indes stich Im mit / deinem vordern Ort zü seiner prüst.

Sticht er dir also zü so nimb Im das ab ywischen baiden henden / In dem gehulß aüf dein lincken seiten Inn dem wind Im deinen vordern Ort deiner mordagst / ober seinem lincken arm zü seiner prüst unnd reiß starckn zü dir trit unnd wind dich damit / von Im zü rückn.

A murder-strike against a thrust to the chest.

Position yourself for this technique as follows: stand with both legs straight and hold the poleax above your head with both hands on the haft and your left arm extended high behind your head (as depicted by the left figure); then, step toward the opponent with your right foot and strike your blade into his head from above.

If he strikes at you thusly from above while your left foot is leading, and you are holding your poleax with your right hand behind your head and your left hand toward him, by the front point (as depicted by the right figure), then parry his strike onto your right side with your ax blade and simultaneously thrust your front point into his chest.

If he thrusts toward you in this same way, then take his strike between your hands and onto your left side, wind your front point over his left arm into his chest, pull strongly toward you, and withdraw while winding.

Poleax 3

Cod. Vinob. 10826, 150r

Duae intorsiones per Securim.

In congressu mutuo, pedem laevum praeponas, manu utraque securim contineas, sinistra hastili adplicata sit iuxta eius laminam, dextra / vero mucroni posteriori, atque mucronem anteriorem intorqueas punctione super brachio sinistro hostis vibrata contra ipsius collum.

Sed / si is idem conetur, te praefigente dextrum pedem, versus sinistrum latus eius impetum avertito securis tuae lamina. Inde autem sinistro consequutus, / nec non laminam securis collo ipsius retrorsum ex parte dextra adplices, et ea ratione adversarium ad te adtrahas.

Verum cum / hostis idem contra te conabitur, tum lamina securis tuae eum impetum excipias in latus tuum dextrum, atque mucrone anteriori visum / hostilem pungas, inde autem ab eo recedas sinistro.

Caeterum si eadem ratione te conpunxerit, consequutus dextro, in latus sinistrum / punctione, eius posteriori mucrone repellito, post dupliciter securi rotata, capiti ipsius, securim infligito, et ita ab hoste recedas.

C94, 165r

Zway einprechen Inn der Mordagst.

Item halt dich also mit disem stuckn Wann Fr zü samen gond / stannd mit deinem lincken fuoß vor unnd halt dein Mordagst Inn baiden henden dein lincken vornen bey / deinem plat dein rechte bey deinem hindern Ort Inn dem prich Im ein mit deinem vordern Ort und stich / Im ober seinem lincken arm zü seinem hals.

Sticht er dir also zü unnd dü mit deinem rechten fuoß gögen Im steest / so nimb Im das ab mit deinem plat aüf dein lincke seiten In dem volg mit deinem lincken schennckel hinnach / unnd seß Im mit deinem plat deiner mordagst hinder seinen hals seiner rechten seiten zeüch Inn damit zü dir.

Ist er dir also einprechen unnd zeüch dich zü Im so nimb Im das ab mit deinem plat deinner mordagst aüf dein / rechte seiten Indes stich Im deinem vordern Ort zü seinem gesicht unnd trit mit deiner lincken schenckel zü / rückn.

Sticht er dir also zü so volg mit deinem rechten schennckel hinnach und nimb Im das ab mit deinem / hindern Ort aüf dein lincke seiten Indes doplier unnd schlag Im ainen mordschlag nach seinem haüpt / trit unnd wind dich damit von Im zü rückn.

Two break-ins with the poleax.

Position yourself thusly for this technique: place your left foot forward and hold your poleax with your left hand on the haft, near the blade, and your right hand by the back point; then, wind over the enemy's left arm and thrust the front point into his throat (as depicted by the left figure).

If he thrusts at you thusly while your right foot is leading, then set his strike onto your left side with your blade, simultaneously pass forward with your left foot, hook your ax blade behind his neck on his right side, and pull him toward you.

If he tries to do the same thing to you, then displace him onto your right side with the blade of your ax, immediately thrust your forward point into his face, and step back with your left foot.

If he thrusts at you with this same technique, then pass forward with your right foot, displace his back point onto your left side, wind your poleax twice, strike his head with a murder-stroke, and withdraw from him.

POLEAX 4

Cod. Vinob. 10826, 150v

Intorsio contra habitum vellendi.

In hoc certamine dextrum praeponas pedem, et in tuta tui defensione securim contineas contra hostem, inde si sinistro fueris consequutus, plaga / letali caput securi concutias.

Sin autem adversarius idem molitur contra te securim tuam utrisque manibus tenentem, dextram anterius / hastili adplicaris, sinistram vero posteriori, atque dextrum pedem praefixeris, versus latus sinistrum capulo repellas, postea intro sequutus / sinistro, mucronem posteriorem supra brachia utraque ingeras, inque ipso actu, mutata sinistra, dextram hostis manum superne securis / lamina contrahas, inferne autem sinistram ipsius abs[1] te brachio sinistro.

Sed si is pari ratione te in latus dextrum tuum diduxerit, dextram / a securi removeas tua, eamque si hostis humero sinistro adplicaris, eum abs te propellas.

Eodem modo si is usus fuerit, pedem / sinistrum referto, atque anteriorem mucronem habitu punctionis in pectus ipsius iniscias, inde autem in tua[2] defensione recedito.

C94, 165v

Ainn Einwinden gögen ainem rezssen.

Item wann Ir zü samen Gond so halt dich also mit disem / stückn stannd mit deinem rechten fuoß vor and halst dein mordagst In guter versaßung / gögen dem man Indes volg mit deinem lincken schennckel hinnach unnd schlag Im ainen mordschlag / nach seinem haüpt.

Schlect er dir also oben zü unnd dü dein mordagst Inn baiden henden hast / dein recht vornen dein lincke bey deinem hinder Ort den rechte schennckel fürgeseßt so nimb Im / das ab mit deinem gehulß aüf dein lincke seiten Inn dem volg mit deinem lincken schennckel hinnach / unnd wind Im deinem hindern Ort ober seine baid Arm verwechsel damit dein lincke hand zü / deinem hindern Ort unnd zeüch Im oben sein rechte hannd mit dem plat und unnden sein lincke / hannd mit deinem lincken Arm zü dir.

Zeücht Er dich also zü Im aüf sein rechte seiten so laß dein / rechte hand gon von deiner mordagst seß Im damit an sein lincke Achsel und scheub in von dir.

Scheubt / er dich also von Im so trit mit deinem lincken schennckel zü rückn unnd wind Im deinem vordern Ort / mit ainem stich zü seiner prüst trit damit von Im zü rückn In güter versaßung.

A wind-in against a pull.

Position yourself in the fight as follows: place your right foot forward and hold your poleax against the opponent in a strong guard, then pass forward with your left foot and strike a murder-stroke to his head.

If he tries to do the same thing to you while you are leading with the right leg, holding the poleax with your right hand at the front of the haft and your left hand by the back point, then set him onto your left side with the haft, pass forward with your left foot, wind your back point over both of his arms, change your left hand, pull his right hand up with your blade, and pull his left hand toward you from below with your left arm.

If he pulls you onto your left side in this way, then release your poleax with the right hand, bring it to his left shoulder, and push him away from you.

If he pushes you away thusly, then bring your left foot back, wind your forward point to thrust into his chest, and withdraw from him in a strong guard.

---

1. The Munich reads *ad*, "toward."
2. The Munich reads *in tuta defensione*, "in a protected guard."

Poleax 5

Cod. Vinob. 10826, 151r

Punctio qua adpetuntur pudenda contra intorsionis habitum.

Ad hanc pugnam ita te componas necesse erit, sinistrum pedem praepones, pro facie tua securim sublatam contineas, dextra hastili medio / sit adplicata, sinistra vero posteriori mucroni, inde si dextro fueris consequutus, pudenda hostis mucrone anteriori haurias.

Sed / si is idem contra te molitus sit, dextrum praefigentem, securimque pro facie tua manu utraque continentem, dextra sit posteriori mucronis / superne vero sinistra iuxta priorem adplicata, tum in latus sinistrum, ipsius impetum mucrone securis tuae posteriori repellas, atque inde / laminam capiti eius infigito quassando.

At si is intra manus ambas capulo suo tuam vim exceperit, consequutus pede laevo, mucronem / posteriorem intra brachium adversariis utrumque contra faciem ipsius intorqueas: At si rursus exceperit, sinistro reducto pede, lamina / securis tuae brachium ipsius dextrum conquassabis, inde securi rotata recedere memineris.

C94, 166r

Einn Gemecht stich gögen ainem einwinden.

Item, schickn dich also Inn diseß stuckn wann Ir zü samen gond stand / mit deinem lincken fuoß vor unnd halst dein mordagst In der höch bey deinem gesicht dein rechte mitten / Inn deinem gehulß dein lincken bey deinem hindern Ort Inn dem trit mit deinem rechten schenckel / hinein unnd stich Im deinem vordern Ort zü seinem gemechten.

Sticht er dir also zü unnd dü mit deinem / rechten fuoß gögen Im steest unnd helst dein mordagst Inn baiden henden bey deinem gesicht dein rechte / unden bey deinem hindern Ort dein lincke oben bey deinem vordern so nimb Im das ab mit deinem hindern / Ort aüf dein lincke seiten Indes haw Im mit deinem plat nach seinem haüpt.

Verseßt er dir das zwichen / baiden hennden Inn sein gehulß so volg mit deinem lincken schenckel hinnach und wind Im deinen / hindern Ort mit ainem stich zwischen seinem baiden Armen hinein zü seinem gesicht verseßt Er dir / das furbaß so trit mit deinem lincken schennckel zü rückn unnd haw Im mit deinem plat nach seinem / rechten Arm wind dich damit von Im zü rückn.

A thrust to the groin against a wind-in.

Position yourself thusly for this fight: place your left foot forward and hold the poleax high in front of your face, with your right hand gripping the middle of the haft and your left hand by the back point; then, pass forward with your right foot and thrust your front point into the enemy's groin (as depicted by the left figure).

If he attempts the same thrust while you are leading with the right foot and holding your poleax in front of your face with both hands, your right at the back point below[1] and your left near the front, then displace his attack onto your left side with your back point and simultaneously strike your blade into his head (as depicted by the right figure).

If he displaces you thusly between both hands on his haft, then pass forward with your left foot and wind your back point between his arms into his face. If he in turn parries, then bring your left foot back, strike your blade into his arm, and wind your poleax into the withdrawal.

---

1. The Latin reads *superne*, "above."

Poleax 6

Cod. Vinob. 10826, 151v

Habitus duo vellendi, unde iactus formatur.

Si palmam hinc ferre volueris, hac ratione te accomodabis, dextrum pedem praeponito, securim sublatam pro facie tua teneto in latere sinistro, / dextra teneat mucronem posteriorem, sinistra autem hastile medium, inde consequitor pede sinistro, atque lamina securis caput hostile / concutito. Sin hostis te repellat inter manum utramque capulo, tum mucronem anteriorem pariter addita lamina sub poplitem dextrum ingeras, / et versum te adtrahas.

Idem vero si is usurparit, tum vicissim et tu laminam, collo ipsius hostis adiungere non dubites ex habitu ictus, / de latere dextro colli, si ex ea effigie firmiter adtraxeris hostem, illius impetus nocere tibi non poterit:

At si adversarius pari te ratione, / adtraxerit superne, in trangulum concedas, et si eum impetum intra manus ambas capulo securis letalis exceperis, nec non sursum / presseris, superne eum infirmum reddideris. Verum inde manum sinistram a securi removeas, eam pedi eius dextro si admoris, atque sublevaris, / supinum hostem prosternes.

C94, 166v

Zway reysen daruß ain wurff geet.

Item wann Fr zü samen Gond so schickn dich also Inn / diseß stuckn stannd mit deinem rechten fuoß vor unnd halt dein mordagst In der hoch bey deinem / gesicht aüf deiner lincken seiten dein rechte hannd vornen bey deinem hindern Ort dein lincke / mitten Inn deinem gehulß Inn dem volg mit deinem lincken schenckel hinnach unnd haw Im / mit deinem plat nach seinem haüpt Nimbst er dir das ab zwischen baiden henden Inn sein gehulß / so wind Im deiner vordern Ort unnd plat under sein rechte kniepug zeuch damit zü dir.

Hat er / dir also unnden angeseßt unnd zeuch dich zü Im so gee Im aüch mit deinem plat deiner mordagst mit / ainem haw umb seiner hals seiner rechten seiten Reiß damit starckn zü dir so mag dir sein anseßen / ain schennckel nit schaden.

Reißt er dich also oben zü Im so trit Inn triangel unnd Nimb Im das / Ab zwischen baiden hennden Inn dein gehulß Truckn damit starckn obersich so nimbstu Im oben das / gewicht Inn dem laß dein lincke hannd von deiner mordagst er greiff damit seinen rechten / schennckel unnd heb In obersich so wurftstu In zü rückn.

Two pulls followed by a throw.

Position yourself for this technique as follows: place your right foot forward and raise the poleax in front of your face on the left side, with your right hand holding the back point and your left hand in the middle of the haft, then pass forward with your left foot and strike your blade into the enemy's head. If he displaces you between his hands on his haft, then wind your front point and blade behind his right knee and pull toward you (as depicted by the left figure).

If he pulls your knee thusly, then strike, hook your blade around his neck on his right side (as depicted by the right figure), and pull him strongly so that his attack cannot harm you.

If he pulls your neck in this same way, then step in a triangle, set his attack aside between your hands on the haft of your murder-ax, strongly press his weapon down so that he is weak, remove your left hand from your ax, grab his right leg, lift up, and throw him onto his back.

Poleax 7

Cod. Vinob. 10826, 152r

Irruptionis forma per Securim.

Si propius ad hostem accesseris, tibiis paribus erectus consistas, securim tuam ambabus manibus sublatam pro facie tua teneas, post / si fueris sinistro ingressus, lamina caput hostis pulsato.

Sed si adversarius idem contra te usurparit, tum eius ictui obviam ferias, / atque eo modo ictum hostis eludas. Verum interea mucronem inferiorem supra brachium ipsius dextrum ingeras, et dextra, adversariis securim / tuae coniungas, inde si te in latus tuum dextrum converteris, intra securim utramque dextram hostis non sine dolore eiusdem coartabis.

Eadem vero ratione si is usus fuerit, tibique extorserit securim, eam hosti concedas, verum manu sinistra exterius compraehendas / ipsius axillam dextram, et firmiter abs te eo habitu adversarium promoveas, sed dextra rursum arripias securim ex eius manibus / extorquendo, atque retrorsum si concesseris, plaga letali caput hostile laminae conflictu saucies.

C94, 167r

Ein einprechen In der mordagst.

Item halt dich also mit Disem stuckn wann Ir zusamen gond stannd mit gleichen fuossen aüfrecht unnd halt dein mordagst Inn baiden henden Inn der hoch vor deinem gesicht Inn dem volg mit deinem lincken schennckel hinnach unnd haw Im mit deinem plat nach seinem haüpt

Hawt er dir also oben zü so haw seinem haw entgögen nimb Im das damit hinweckn inn dem prich Im ein mit deinem undern Ort ober seinen rechten Arm unnd erwisch Im mit deiner rechten hannd sein mordagst zusampt der deinem wend dich damit aüf dein rechte seiten so trucknstu Inn sein rechte hannd zwischen baiden Mordagsten

Ist er dir also ein geprechen unnd begert dir dein Mordagst zü nemen so laß Im die Inn dem greiff Im mit deiner lincken hand aüffwendig under sein recht Achsen truckn In damit starkcn von dir unnd mit deiner rechten greiff widerumb In dein mordagst Indes reiß die aüß seiner hand trit damit zü rückn unnd haw Im ainem mordschraich mit deinem plat nach seinem haüpt.

A break-in with the poleax.

As you approach the enemy, stand with both legs straight and hold your poleax in the high guard in front of your face, then pass forward with your left foot and strike your blade into his head.

If he strikes at you from above with this same technique, then counter him by striking against his blow, take him aside, thrust your lower point over his right arm, grab both his haft and yours with your right hand, and turn yourself onto your right side, thus smashing his right hand between the two poleaxes (as depicted by the right figure).

If he tries to disarm you using this same technique, then let go of your poleax, bring your left hand under his right armpit, on the outside, and push him away from you; at the same time, grab your poleax with your right hand, pull it out of his hand, step back, and strike your blade into his head with a murder-blow.

Poleax 8

Cod. Vinob. 10826, 152v

Plaga letalis contra habitum vellendi.

Sic te in hoc certaminis genus adaptes, dextrum pedem praepones, hastile securis utraque manu contineas, atque plagam letalem capiti adversariis / infligas lamina tuae securis.

Sed si is idem fuerit conatus contra te sinistrum praefigentem, tum mucronem anteriorem inter brachium / utrunque hostis iniscias, et si collo eius ex parte dextra adplicaris, ipsius ictum mucrone posteriori repellas, inde laminae curvatura / hostem ad te contrahas.

At si is pari contra te ratione fuerit usus, tum dextram commutes ab interius, exterius in securi, eoque / habitu hostis securim deorsum deflectas supprimendo, nec non latus colli hostilis sinistrum lamina proscindas.

Sin vero idem adversarius / usurparit, eam vim securi versus dextrum latus tuum repellas, verum interea visum ipsius anteriori mucrone si pupugeris, ab / hoste recedas.

C94, 167v

Ain mordschlag gögen ainem reyssen.

Item wann Ir zü samen Gond so halt dich also In disem / stuckn stannd mit deinem rechten fuoß vor unnd halt dein mordagst mit baiden henden Inn deinem gehulß / Inn dem schlag Im ainem Mordschlag mit deinem plat nach seinem haüpt.

Hawt er dir also oben / zü unnd dü mit deinem lincken fuoß gögen Im steest so gee Im mit deinem vordern Ort zwischen seinen / baiden armen hinnein seß Im damit an seinen hals seiner rechten seiten unnd nimb Im seinem / straich mit deinem hindern Ort hinweckn Indes Reiß in mit deinem hacken deiner plat zü dir.

Reist / er dich also zü Im so verwechsel dein rechte hand von Im wendig aüß wendig widerumb In dein mordagst / wind damit sein mordagst undersich unnd schnied Im mit deinem Blat nach seinem hals seiner lincken / seiten.

Hat er dir also angeseßt so nimb Im das ab mit deiner mordagst aüf dein rechte seiten in dem stich / Im mit deinem vordern Ort zü seinem gesicht trit unnd wind dich damit von Im zü rückn.

Poleax

A murder-stroke against a pull.

Position yourself for this fight as follows: place your right foot forward, hold the haft of your poleax with both hands, and strike a murder-stroke to your opponent's head with your blade (as depicted by the right figure).

If he attempts this same murder-stroke while your left foot is leading, then strike your front point between his arms, hook his neck from the right side (as depicted by the left figure), and parry his strike with your back point as you pull him toward you with the curved part of your blade.

If he pulls you thusly to him, then quickly switch your right hand from the inside to the outside of your poleax, wind over his weapon, drive it downward, and slice your blade into the left side of his neck.

If he tries to do the same thing to you, then set his attack onto your right side with the poleax, immediately thrust your front point into his face, and wind yourself into the withdrawal.

Poleax 9

Cod. Vinob. 10826, 153r[1]

Ictus supernus contra punctionem.

In mutuo congressu hoc sinistrum praeponito pedem, atque pro facie tua securim sublatam utraque manu teneas, inde dextro insequutus, / lamina tuae securis caput hostis quassabis.

Sed si idem is contra te molitus fuerit sinistrum praeponentem, securis mucronem anteriorem / sinistra manu tenentem contra hostem, dextra autem posteriorem, capulo intra manum utramque ictum hostilem excipito, interea / mucrone anteriori ipsius fodito collum.

Sin autem is te pari ratione fuerit adgressus, tum eum impetum si in latus tuum dextrum / repuleris mucrone anteriori super brachio sinistro directo faciem hostilem pungito. Sin vero securi eum conatum removerit, tum / posteriorem cuspidem securis lateri ipsius sinistro infigas.

Idem cum is usurparit, lamina eum impetum excipito, post dextro consequutus, / pectori hostis mucronem anteriorem inter brachium eius utrumque adigas, atque ab eo rotata securi recedas.

C94, 168r

Ein oberhaw gögen ainem stich.

Item schickn dich also In Diseß stuckn wann Ir zü samen gond / stand mit deinem lincken fuoß vor unnd halt dein mordagst in baiden henden Inn der hoch vor deinem / gesicht Inn dem volg mit deinnem rechten schennckel hinnach unnd haw Im mit deinem plat nach seinem / haüpt.

Hawt Er dir also oben zü unnd dü mit deinem lincken fuoß gögen Im steest unnd halst dein / mordagst mit deiner lincken hand bey deinem vordern Ort gögen dem man dein rechte bey deinem / hindern so nimb Im seinem haw zwischen baiden handen Inn deinem gehulß hinweckn Inn dem stich / Im mit deinem vordern Ort zü seinem hals.

Sticht er dir also zü so seß Im das ab aüf dein rechte seiten / In dem stich Im mit deinem vordern Orth ober seinem lincken Arm zü seinem gesicht verseßt er dir das / mit seiner mordagst so wind Im deinem hindern Ort nach seiner lincken seiten.

Windt er dir also unden / ein so seß Im das ab mit deinem plat (in dem [crossed out]) Inn dem volg mit deinem rechten schenckel hinnach und stich / Im deinen vordern Ort zwischen seinen baiden armen zü seiner prüst trit unnd wind dich damit von / Im zü rückn.

Poleax

An *oberhaw*[2] against a thrust.

Position yourself thusly in this fight: place your left foot forward and hold the poleax high front of your face with both hands; then, pass forward with your right foot and strike your blade into the enemy's head (as depicted by the left figure).

If he strikes an *oberhaw* at you while your left foot is leading and you are holding the poleax with your left hand by the front point, toward the opponent, and your right hand at the back, then displace his strike between both hands on the haft (as depicted by the right figure), and simultaneously thrust your front point into his throat.

If he thrusts thusly toward you, then set him onto your right side while thrusting your front point over his left arm into his face. If he displaces your thrust with his poleax, then wind your back point into his left flank.

If he winds you in this same way, then set him aside with your blade, pass forward with your right foot,[3] thrust your front point between his hands into his chest, and wind yourself into the withdrawal.

---

1. This plate is misnumbered "152."
2. German, "over strike; strike from the upper openings." See Latin *ictus superus*, "above strike."
3. The Latin reads "left foot."

Poleax 10

Cod. Vinob. 10826, 153v

Habitus seu Castrum, ubi securim pectori adiunctum habes.

In hoc certamen hac ratione componas te, sinistrum pedem praeponas, manibus utrisque securim contineas, contraque hostem sinistram / hastilis parti anteriori adplicatam, dextram vero posteriori mucronis pectori coniunctam et admotam, inde dextro pede consequitor, et / ex habitu ictus, qui a vento adpellationem fortitus est de pectore tuo securim versus caput hostis dirigas.

Verum si is te eadem ratione / fuerit adgressus dextrum praefigentem, dextra securis partem posteriorem continentem contra hostem, sinistra vero anteriorem in / latere tuo sinistro, tum dextro pede relato, capulo intra manum utramque impetum hostis excipias, inde autem rursus dextrum pedem intro / collocabis, atque mucronem posteriorem in faciem ipsius dirigas.

Sin id inter ambas manus securi exceperit adversarius, sinistro insequutus, / laminam letaliter capiti ipsius infligito, et ab hoste tuto curabis recedere rotata securi.

C94, 168v

Ainn prustleger gögen ainem abnemen.

Item wann Ir zü samen Gond so shickn dich also Inn / diseß stuckn stannd mit deinem lincken fuoß vor unnd halt dein mordagst In baiden henden dein linke / bey deinem vorden Ort gögen dem man dein rechte unnden bey deiner hindern Ort ob deinner / prüst Inn dem volg mit deinem rechten schennckel hinnach und wind Im mit ainem windthaw von / deiner prüst zü seinem haüpt.

Hawt Er dir also oben zü unnd dü mit deinem rechten fuoß gögen Im / steest unnd helst dein mordagst mit deiner rechten hand bey deinem hinder Ort gögen dem mann / dein lincken bey deinem vordern aüf deiner lincken seiten so trit mit deinem rechten schennckel zü rückn / unnd verseßt Im den haw zwischen baiden henden Inn dein gehulß In dem trit mit deinem rechten schennckel / wider hinein und wind Im deinen hindern Ort zü seinem gesicht.

Verseßt er dir das zwischen baiden / henden In sein mordagst so volg mit deinem lincken schenckel hinach unnd haw Im ainem mordstreich / mit deinem plat nach seinem haüpt trit unnd wind dich damit von Im zü rückn.

A guard at the chest against an *abnemen*.[1]

Position yourself as follows in this fight: place your left foot forward and hold your poleax with your left hand near the front point, toward the enemy, and your right hand at the back point, by your chest (as depicted by the right figure); then, pass forward with your right foot and strike a *windthaw*[2] from your chest to his head.

If he strikes you thusly while you are leading with the right foot and holding the back point toward him with your right hand, the front point on your left side with your left hand (as depicted by the left figure), then bring your right foot back, displace his *windthaw* between your hands on the haft, step forward again with your right foot, and wind your back point into his face.

If the displaces you between his hands in this way, then pass forward with your left foot, strike your blade into his head with a murder-stroke, and wind yourself into the withdrawal.

---

1. German, "taking-off." See Latin *habitus aversionis*, "parry; setting-aside."
2. German, "wind strike."

POLEAX II

Cod. Vinob. 10826, 154r[1]

Duae mucronum posteriorum intorsiones.

Dextrum pedem in mutuo hoc congressu praeponas, securim manu utraque contineas, dextra sit mucroni posteriori adplicata contra hostem, / sinistra vero iuxta laminam, ex ea igitur constitutione sinistro si fueris insequutus pede, collum hostis mucrone posteriori / intra brachium utrumque inserto figas.

Sed si adversarius idem contra te usurparit, tibiis paribus consistentem, nec non pro facie tua / securim sublatam mediam tenentem utrisque manibus, tum versus dextrum latus tuum mucrone posteriori hostilem impetum repellas. Postea / sinistro consequitor, atque mucronem posteriorem contra faciem adversariis inter brachium ipsius utrumque inferas. Sin vero eum conatum / repulerit, derepente laminam pro viribus capiti eius infligas.

At si caput tuum pari ratione hostis adgressus fuerit, sinistrum pedem / referas, capuloque inter manum tuam utramque eam vim excipias, inde lamina tuae securis, adversariis securim deorsum supprimas, et si / anteriori mucrone visum ipsius punxeris, ab hoste recedas.

C94, 169r

Zway Einwinden mit Fren hindern Ortern.

Item halt dich also mit Disem stuckn wann Fr zü samen / gond stannd mit deinem rechten fuoß vor unnd halt dein mordagst Inn baiden henden dein rechten bey deinem hindern / Ort gögen dem mann dein lincke hinder bey deinem plat In dem volg mit deinem lincken schenckel hinach / unnd wind Im deinem hindern ort zwischen baiden Armen hinein zü seinem hals.

Windt er dir also ein unnd / dü mit gleichen füossen aüf recht gögen Im steest unnd helst dein mordagst mit baiden henden mitten Inn deinem gehulß / Inn der hoch vor deinnem gesicht so nimb Im das ab mit deinem hindern Ort aüf dein rechte seiten In dem / volg mit deinem lincken schenckel hinnach unnd wind Im deinem hindern ort zwischen baiden Armen hinein / zü seinem gesicht Verseßt er dir das so haw Im behend ainem mordstraich mit deinem plat nach seinem haüpt.

Hawt er dir also oben zü so trit mit deinem lincken schennckel zü rückn und verseß Im das zwischen baiden henden / Inn dein gehulß Inn dem reiß Im mit deinem Blat sein mordasgt unndersich und stich Im mit deinem / vordern Ort zü seinem gesicht unnd wind dich damit von Im zü rückn.

Two wind-ins using the back point.

Position yourself for this technique as follows: place your right foot forward and hold the poleax with your right hand at the back point, toward the enemy, and your left hand near the blade (as depicted by the left figure); then, pass forward with your left foot, wind your back point between his arms, and thrust into his throat.

If he employs this same wind against you while you are standing with your legs even, holding the middle of your haft with both hands raised in front of your face (as depicted by the right figure), then set his attack onto your right side with your back point, pass forward with your left foot, and wind your back point between his arms into his face. If he parries you, then immediately strike your blade into his head with a powerful murder-stroke.

If he strikes at your head thusly, then bring your left foot back, parry his attack between both hands on your haft, press his poleax down with your blade, thrust your front point into his eyes, and wind yourself into the withdrawal.

---

1. This plate is misnumbered "153."

Poleax 12

Cod. Vinob. 10826, 154v

Punctio unde pectus adpetitur, contra ictum medium.

Si mutuo congredimini, tu ea ratione te componas contra hostem, pedem sinistrum praefigere non dubites, securim sinistra teneas / iuxta laminam versus adversarium, dextra autem sit mucronis posteriori adplicata secundum caput tuum, post insequutus dextro, mucrone / anteriori pectus fodias.

At si adversarius pari ratione te fuerit superne adgressus dextrum praeponentem, ex latere sinistro / securim manibus ambabus continentem, sursum mucronem posteriorem tollas, et punctionem hostilem repellas in latus sinistrum tuum, / inde autem consequutus sinistro pede, latus adversariis dextrum lamina ex habitu mediis ictus concutias.

Idem vero si hostis molitus / fuerit, securis tuae capulo eum impetum excipias, interea lamina eiusdem, hostis securim supprimas, in latus sinistrum tuum, atque / pectus ipsius pungas. At si id inter manum utramque capulo exceperit, lamina confestim caput hostile ferias, et inde ab eo recedas.

C94, 169v

Ainn prüst stich gögen ainem mittlehaw.

Item wann Fr. zü samen Gond so halt dich also mit disem / stuckn stannd mit deinem lincken fuoß vor unnd halt dein mordagst mit deiner lincken hand bey deinem / plat gögen dem mann dein rechte bey deinem hindern Ort hinder deinem haüpt In dem volg mit / deinem rechten schennckel hinnach unnd stich Im mit deinem vordern ort zü seiner Brust.

Sticht er also von / oben aüf dich unnd dü mit deinem rechten fuoß gögen Im steest unnd helst dein mordagst In baiden henden / aüf deiner lincken seiten so gee mit deinem hinder Ort aüf unnd nimb Im seinem stich hinweckn aüf dein / lincke seiten Inn dem volg mit deinem lincken schennckel hinnach und haw Im ainem Mittlehaw nach / seiner rechten seiten mit deinem plat.

Hawt er dir also nach deiner rechte seiten so nimb Im das ab mit deinnem / gehulß deiner mordagst Indes reiß mit deinem plat sein mordagst aüf dein lincke seiten undersich und / stich Im nach seiner prüst Verseßt er dir das zwischen baiden henden In das gehulß so haw Im behend mit deinem / plat nach seinem haüpt unnd trit damit von Im zü rückn.

A thrust to the chest against a *mittelhaw*.[1]

When you approach each other at the same time, position yourself as follows: place your left foot forward and hold the poleax with your left hand near the blade, toward the adversary, and your right hand at the back point, behind your head (as depicted by the left figure); then, pass forward with your right hand and stab your front point into his pectoral.

If he thrusts at you thusly while you are leading with the right foot and holding your poleax on your left side with both hands (as depicted by the right figure), then raise your back point up and set his thrust onto your left side; at the same time, pass forward with your left foot and strike a *mittelhaw* with your blade into his right flank.

If he strikes this same *mittelhaw* to your flank, then take him aside with the haft of your poleax, push his poleax down onto your left side with your blade, and thrust into his chest. If he parries your thrust between his hands on the haft, then immediately strike your blade into his head, and withdraw from him.

---

1. German, "middle-strike"

Poleax 13

Cod. Vinob. 10826, 155r

Aversionis modus contra eum ictum, quo pes sauciatur.

Si in hostis conspectum accesseris, pedibus paribus rectus consistas, supraque caput securim sublatam contineas utrisque manibus, ex / eo igitur corporis positu intro procedas sinistro, et ex alto ictu directo lamina pedem eius praefixum verberes.

Sin autem laevum praeposueris, / nec non istum pedem adpetiverit, securim demittas, et sinistro pedi eam praeponas, mucro anterior sit humo inclinatus, eoque habitu ictum / ipsius avertas. Post sublata securi, dextro pede insequutus, mucronem anteriorem collo ipsius infigas. Sed si eundem impetum / lamina in latus dextrum removerit, inferne mucronem transmittas, atque latus sinistrum fodias.

Idem autem si is moliatur, lamina tuae / securis in latus sinistrum repellas, inde dextro consequutus, utrumque mucronem inter brachium utrumque contra hostis faciem vel pectus / dirigas, postea autem lamina caput eius si quassaris, ab eo recedas.

C94, 170r

Ein abnemen gögen ainem haw zü dem Schennckel.

Item schickn dich also Inn diseß stuckn wann Fr zü samen gond / stand mit gleichen fuossen auf recht unnd halt dein mordagst In baiden henden In der hoch ob deinem haupt In / dem trit mit deinem lincken schennckel hinein unnd haw Im von oben nider mit deinem plat nach seinem furgeseßten / schennckel.

Steestu dann also mit deinem lincken fuoß gögen Im unnd er dir darnach schlecht so far / mit deiner mordagst unndersich unnd seß dir für deinen lincken schennckel deinen vordern Ort auf die erden / nimb Im damit seinem straich ab Inn dem gee auf mit deiner mordagst unnd volg mit deinem rechten schenckel / hinnach stich Im damit deinen vordern Ort zü seinem hals nimbst er dir das ab mit seinem plat auf sein recht / seiten so wechsel Indes behend unnden durch unnd stich Im nach seiner lincken seiten.

Sticht er dir also zü so nimb Im / das ab mit deinem plat auf dein lincken seiten In dem volg mit deinem rechten schennckel hinnach und wind Im / deinem hindern unnd vordern Ort zwischen baiden Armen hinein zü seinem gesicht oder der prust Indes haw Im / mit deinem plat nach seinem haupt und trit damit von Im zü rückn.

An *abnemen*[1] against a strike to the leg.

Position yourself during the approach as follows: stand up straight, with your feet even, and hold the poleax above your head with both hands; then, advance with your left foot and strike low to his leading leg (as depicted by the right figure).

If you are leading with your left foot and he strikes at you thusly, then lower your front point to the ground in front of your left foot, and parry his strike (as depicted by the left figure); then, raise your poleax, pass forward with your right foot, and thrust your front point into his throat. If he sets you onto his right side with his blade, then quickly change under him and stab into his left flank.

If he tries the same thing, then set his thrust onto your left side with your blade, pass forward with your right foot, wind both of your points between his arms into his face or chest, strike your blade into his head, and withdraw from him.

---

1. German, "taking-off." See Latin *habitus aversionis*, "parry; setting-aside."

Poleax 14

Cod. Vinob. 10826, 155v

Cancellatae aversionis habitus contra punctionem, qua pubes adpetitur.

Hac ratione te accomodes, dextrum praefigas pedem, securim supra caput sublatam utraque manu teneas, dextram mucroni anteriori / contra hostem adplices, sinistram vero posteriori, inde sinistro consequutus, haurias adversariis pudenda mucrone anteriori.

Idem vero si is contra te fuerit molitus, sinistrum praeponentem, nec non securim tenueris cancellatis brachiis adversus hostem, sinistra / manus iuxta securis laminam sit posita, dextra anterius capulo adplicata, sinistro pede reducto, ipsius punctionem mucrone posteriori / in latus sinistrum tuum removeas, interea confestim sinistro rursus progreditor, atque anteriorem mucronem ab inferne sursum / in faciem vel hostis pectus dirigas.

Adversario pari ratione utente, capulo id intra manum tuam utramque excipito, inde in triangulum / si concesseris, lamina caput eius sauciato, atque ab hoste eo modo usus recedito.

C94, 170v

Ain gschrenckts abnemen gogen ainem gemecht stich.

Item wann Ir zü samen gond so schickn dich also / Inn diseß stuckn stannd mit deinem rechten füoß vor unnd halt dein mordagst In baiden henden ob deinem / haüpt dien rechte bey deinem vordern Ort gögen dem man die linct bey deinem hindern In dem volg / mit deinem lincken schennckel hinach unnd stich Im mit deinem vordern Ort zü seinem gemechten.

Sticht er dir also zü deinen gemechten unnd dü mit deinem lincken füoß gögen Im steest und helst dein mordagst / mit geshrennckten Armen gögen dem man dein lincke hand bey deinem plat die recht vornen Inn / dem gehulß so trit mit deinem lincken schennckel zü rückn und nimb Im seinen stich mit deinnem / hindern Ort hinweckn aüf dein lincke seiten Indes trit behend mit deinem lincken schenckel wider / hinein unnd wind Im deiner vordern Ort von unnder aüf zü seinem gesicht oder der prust.

Windt / er dir also ein so nimb Im das ab zwischen baiden henden In dein gehulß In dem trit In trangel unnd / haw Im mit deinem plat nach seinem haüpt zeüch dich damit von Im zü rückn.

A crossed *abnemen*[1] against a thrust to the groin.

Position yourself as follows: place your right foot forward and hold the poleax above your head with your right hand by the front point, toward the enemy, and your left hand at the back point (as depicted by the right figure); then, pass forward with your left foot and thrust your front point into his groin.

If he thrusts thusly at your groin while you are leading with the left foot and holding your poleax with crossed arms, your left hand near the blade, your right hand forward on the haft (as depicted by the left figure), then bring your left foot back, set his thrust onto your left side with the back point, quickly step forward again with your left foot, and wind your front point from below up into his face or chest.

If he attempts this same wind, then parry him between your hands on the haft, step in a triangle, strike your blade into his head, and withdraw from him.

---

1. German, "taking-off." See Latin *habitus aversionis*, "parry; setting-aside."

Poleax 15

Cod. Vinob. 10826, 156r[1]

Ictus letalis contra impositionem cancellatam, ut infra patebit.

In mutuo congressu sic te contra hostem accommodes necesse erit, pedem sinistrum praeponere non dubites, brachiisque instar crucis / constitutis securim teneas contra adversarium, inde habitu punctionis securis mucronem anteriorem, brachio hostis sinistro adplices.

Sin autem contra eum sinistrum praefigens consistas, sinistrumque brachium pungere is molitur, sinistro retracto, versus sinistrum latus tuum / punctionem hostilem lamina repelles: Verum celeriter rursus ingreditor, atque ictum letalem capiti ipsius lamina inferas.

Sin vero / is pari ratione te adgreditur, eam vim capulo intra manum utramque excipito, inde autem consequutus pede dextro, mucronem anteriorem / versus faciem adversariis vel pectus intendito. Id si exceperit, dextro pede rursum reducto, ictu, qui a vento adpellationem / indeptus est, sinistrum pedem eius praefixum si sauciaris, ab hoste recedas

C94, 171r

Ein mordthaw gögen ainem geschrenckten anseßsen.

Item halt dich also mit disem stuckn wann Fr zü samen / gond stannd mit deinem lincken füoß vor unnd halt dein mordagst mit geschrencken Armen / gögen dem man Inn dem seß Im mit deinem vordern Ort an seinem lincken arm mit ainem stich.

Stastu dann aüch also mit deinem lincken füoß gögen Im unnd er dir nach deinem lincken arm sticht so / trit mit deinem lincken schennckel zü rückn unnd nimb Im das ab mit deinem plat auf dein lincken seitten / Indes trit behend wider hinein unnd haw Im ainem mordhaw mit deinem plat nach seinem haüpt.

Hawt er dir also oben zü so nimb Im das ab zwischen baiden henden In dein gehulß Im dem volg mit deinem / rechten schennckel hinnach und wind Im deinen vordern ort mit ainem stich zü seiner gesicht / oder der prüst verseßt Er dir das so trit mit deinem rechten schennckel wider zü rückn und haw Im / ainem windthaw nach seinem lincken furgesetzen schenckel trit damit von Im zü rückn.

A murder-stroke against a crossed *ansetzen*.[2]

Position yourself thusly as you approach the enemy: place your left foot forward and hold your poleax with crossed arms; then, press your front point onto his left arm (as depicted by the right figure).

If you are standing with your left leg toward him and he thrusts at your left arm, then bring your left foot back, set his thrust onto your left side with your blade, quickly step forward again, and strike your blade into his head with a murder-stroke (as depicted by the left figure).

If he strikes a murder-stroke at you, then parry him between your hands on the haft, pass forward with your right foot, and thrust your front point into his face or chest. If he parries you, then bring your right foot back, strike a *windthaw*[3] into his leading left leg, and withdraw from him.

---

1. This plate is misnumbered "155."
2. German, "setting-on." See Latin *impositio*, "press."
3. German, "wind strike."

POLEAX 16

Cod. Vinob. 10826, 156v

Gemina impositio securis.

Ita compones te in hoc certamen, dextrum pedem praeponito, manum utramque mucroni posteriori securis adplicato supra caput allevatae, inde sinistro consequutus, latus capitis hostilis dextrum transversario ferito ictu.

Sed si idem is contra te moliatur, sinistrum / contra se praeponentem in habitu librae, et securim manu utraque contineas ita, ut anterior mucro terrae nitatur, securi sublata, / capulo securis tuae, intra utramque manum tuam, hostis ictum transversarium repellas.

Sin vero eadem ratione tu ab hoste repulsus / fueris, tum mucronem securis anteriorem adplica brachio ipsius sinistro.

Verum si idem is contra te usurparit, lamina id repellas, /et interim securim hostilem lamina tuae securis intercludas, et si pro viribus mucronem anteriorem corpori seu thoraci eius / infixeris, ruere adversarium ea punctione compelles.

C94, 171v

Zway anseßen Inn der mordagst.

Item wann Fr zp samen gond so hat dich also / mit disem stuckn stannd mit deinem rechten füoß vor unnd halt dein mordagst Inn baiden henden bey deinem / hindern Ort Inn der hoch ob deinem haüpt Inn dem volg mit deinem lincken schennckel hinnach und haw / Im ainen zwirchhaw nach seinner rechten seiten seines haupt.

Hawt er dir also oben zu unnd dü mit deinem lincken / fuoß gögen Im steest, wol Inn der wag, dein Mordagst Inn baiden henden dein vordern Ort gögen der erden so gee / aüf mit deiner Mordagst unnd nimb Im seiner zwirchhaw zwischen baiden henden Inn dein gehulß hinweckn.

Hat er dir das also ab genemen so seß Im mit deinem vordern Ort deiner mordagst mit ainem stich an seinen / lincken arm.

Hat er dir also angeseßt so nimb Im das ab mit deinem plat Indes spery Im sein mordagst mit deinem / plat unnd stich Im deiner vordern Ort mit ainem nach truckn mit sterckn In seinen leib, so stichstu Im zü / der erden.

Two *ansetzen*[1] using the poleax.

Position yourself like so in this fight: place your right foot forward and hold the poleax high above your head, with both hands by the back point; then, pass forward with your left foot and strike a *zwerchaw*[2] to the right side of your enemy's head.

If he strikes a *zwerchaw* at you while you are standing in a balanced position with your left forward, holding your poleax in both hands with the front point tilted toward the ground, then raise your weapon and parry his *zwerchaw* between your hands on the haft.

If he parries you thusly, then press your front point to his left arm with a thrust.

If he does the same thing to you, then simultaneously set his thrust aside and trap his weapon with your blade, drive your front point toward his upper body with all your might (as depicted by the left figure), and force him to fall over.

---

1. German, "setting-on." See Latin *impositio*, "press."
2. German, "twitch strike." See Latin *ictus transversarius*, "transverse strike; strike from side to side."

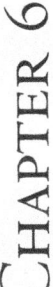

Chapter 6

Various Weapons

"**VARIA** arma contra variis armis utentes, ut cum quis ex improviso obruit, / se possit defendere, sunt autem formae eorum, 12" (Cod. Icon. 393, 14v, 8-9).

"**VARIOUS** weapons employed against various weapons, so that when someone overwhelms [you] unexpectedly, it will be possible to defend yourself. There are 12 images of them."

In its entirety, this chapter considers a dozen possible weapon combinations. We have included 10 of them in this book, namely, every instance in which a polearm is used against another weapon.[1] Two of the polearms—boar spear and javelin—are not depicted elsewhere.

The boar spear appears in three folia. It is identified by the Latin term *venabulum*, "hunting spear," and the German term *schwein spiess*, "boar spear." It is roughly 8 feet in length, with a 12- to 18-inch leaf-shaped blade. In two folia, a pair of lugs protrude from the base of the spearhead, an invention designed to prevent an impaled boar from charging up along the shaft and goring the hunter in its final throes. In the third, the blade is shaped like an arrowhead, and tufts of cloth or feathers appear at the base instead of lugs, an unconventional design for a boar-hunting spear of this period.

The weapon that appears to be a javelin is named by the Latin term *hastula*, which literally means "short spear," thereby supporting our hypothesis that the shortstaff chapter in fact deals with spear techniques. It consists of a diamond-shaped blade mounted on a wooden shaft roughly 5 to 6 feet in length. We conclude that the weapon shown here is the actual short spear for which the shortstaff is used to train.

Three of the other weapons—longsword, dussack, and rapier—are covered in other sections of the *Opus*, as well as other manuals, and will not be discussed here.

The graphic depictions of gore in this chapter set it apart from the rest of the manual, which clearly shows combatants in a training environment. It is also interesting to note that Mair does not rank the weapons in terms of their advantage against one another, unlike, for instance, George Silver in his 1617 *Paradoxes of Defence*.

---

1. One folio from this chapter deals exclusively with halberds and was thus placed in the "Halberd" chapter of this book.

Various 1

Cod. Vinob. 10826, 15r

Forma pugnandi hastula, qua peregrinantes utuntur contra ensem longum.

In hastulae habitum hac ratione te accommodabis, laevum pedem praeponas, in latere dextro eam manu utraque contineas, mucronem / anteriorem contra hostem porrectum. Sin autem et is sinistro praefixo versus te constiterit, atque in habitu ictus superni ensem levarit, / caputque ferire molitur, tum celeriter mucronem hastulae, hostis corpori impellito.

Verum si is idem contra te conetur, laevum / pedem reducto, eiusque hastam transversario ferias in latere adversariis sinistro, ea itaque ratione eius impetum repelles.

Eo / autem idem contra te usurpante, conatumque tuum impediente, insequutus pede dextro, mucrone hastae anteriori hostis corporis / latus dextrum si fixeris, ab eo recedere poteris.

C93, 227r

Das schefftlin gögen dem schwert.

Item schickn dich also In Diseß sutckn mit deinem schefflin / stannd mit deinem lincken vor unnd hallts Inn baiden henden aüff deiner rechten seiten deinen / vordern Ort oder spieß gögen dem amn stat er dann aüch mit sener lincken fuoß gögen dir unnd helst / sein swert Inn ainem oberhaw gögen dir Inn der hoch unnd begert dir nach deinem haüpt zü hawen so / stich Im behend mit deinem vordern Ort deines schefftlins zp seinem leib.

Sticht er dir also zü so trit mit / deinem lincken schennckel zü rückn unnd ahw Im mit deinem swert nach seiner stanngen seines schefftlins / aüf sein lincken seten mit der zwirch so minbstu Im seinem stich damit hinweckn.

Hawt er dir also / entgögen unnd nimpt dir deinem stich hinweckn so volg mit deinem rechten schennckel hinnach unnd stich / Im mit deinem vordern Ort deines schefflins nach seiner rechten seiten leibs trit damit von Im / zü rückn.

The short spear against the longsword.

Position yourself for this technique as follows: place your left foot forward and hold the short spear on your left side with both hands, its front point extended toward the enemy (as depicted by the left figure). If he then stands against you with his left foot leading, raising his sword in the guard for an *oberhaw*[1] to your head (as depicted by the right figure), then quickly thrust your front point into his body.

If he thrusts at you thusly, then bring your left foot back and strike the shaft of his spear on his left side with a *zwerchaw*,[2] so that you set his thrust aside.

If he employs this same *zwerchaw* against you, thereby setting your thrust aside, then pass forward with your right foot, thrust your front point into his right flank, and withdraw from him.

---

1. German, "over strike; strike from the upper openings." See Latin *ictus superus*, "above strike."
2. German, "twitch strike." See Latin *ictus transversarius*, "transverse strike; strike from side to side."

Various 2

Cod. Vinob. 10826, 15v

Bipennis contra gladium longum.

Ad habitum regendae bipennis, si rite et more athletarum te componere voles contra hostem gladio utentem, pedem sinistrum / praefigere memineris, iuxta mucronem posteriorem bipennem manu dextra contra hostem contineas, medio bipennis sinistra / sit applicata recta pro facie tua sursum, ex latere tuo sinistro, ea igitur ratione confestim caput adversariis ferias.

Verum / si hostis superne te adgreditur modo consimilis praeponentem itidem sinistrum, ensemque super humero dextro continentem, manu sinistra / ensis nodum versus hostem, acie longa praemissa si eius bipennem concusseris, ictus ipsius irritus apparebit. Caeterum si ex eo / habitu in triangulum statim concesseris, caput eius ex latere sinistro aciem longam praemittendo quassabis.

Verum si is eadem contra / te moliatur, gemino passu instando consequutus, superne et inferne contra ipsum certabis pungendo dupliciter, atque inde / ab hoste recedas.

C93, 227v

Die Hellenpart gögen dem swert.

Item wan Fr zü samen Gond so schickn dich also Inn diseß / stuckn mit deiner hellenparten stannd mit deinem lincken füoß vor unnd hallt dein hellenparten mit / deiner rechten hannd bey denem hindern Ort gögen dem man dein lincke Inn der mitt gestrakcts vor / deinem gesicht Inn der hoch aüf deinner lincken seiten Indes haw Im behend nach seinem haüpt.

Hawt er dir / also oben zü unnd dü aüch mit deinem lincken füoß gögen Im steest unnd helst dein swert aüf deiner rechten / Achsel den knopff Inn der lincken hand gögen dem man so haw Im mit lannger schneid nach seinner / hellenparten so nimbstu Im seinen haw damit hinweckn Indes trit Inñ Trianngle unnd haw Im mit langer / schneid nach seinem haüpt seiner lincken seiten.

Hawt er dir also oben zü so trit zwifach hinnein mit ainem / nachraisen unnd stich Im unndern unnd oben zwifach zü trit damit von Im zü rückn.

The halberd against the longsword.

Take position with your halberd against an enemy wielding a longsword as follows: place your left foot forward and hold the halberd in a high guard with your right hand near the back point, toward the enemy, and your left hand gripping the middle in front of your face on your left side (as depicted by the left figure); then, quickly strike from this position into his head.

If he strikes at your head in this same way while you are leading with the left foot and holding your longsword above the right shoulder, the pommel in your left hand, toward the opponent (as depicted by the right figure), then strike his halberd with your long edge to set him aside, immediately step in a triangle, and strike your long edge into the left side of his head.

If he attempts to strike thusly at your head, then pursue him[1] with a double-step forward, thrusting twice toward him from above and below, and then withdraw from him.

---

1. Note that here Mair uses the German term *nachraisen*, "traveling after," i.e. striking after the opponent commits to his attack.

Various 3

Cod. Vinob. 10826, 13r

Lancea contra ensem.

In congressu ad hostem sic te praeparabis, laevum pedem praefiges, lanceam anterius manu sinistra contra adversarium contineas, dextram / Lanceae adplicabis retro pedem dextrum in latere dextro, atque ex eo corporis habitu visum eius adpetito.

Sed si is idem usurparit / contra te, pedem dextrum praeponentem, et in habitu mutatoriis ensis tuus in latere sinistro constiterit, mucro terrae inclinatus, si / eum laevaris, lanceamque ipsius concusseris, impetus adversariis ea ratione in latere tuo dextro repelletur.

Verum si tuum lanceam / eo modo conquassarit ictu ensis sui, lanceam deorsum dirigito, sed confestim eam rursus tolles, in medio Lanceam strenue continuas, / atque latus corporis adversariis sinistrum transfigas.

Sin vero te eadem ratione fuerit adgressus, tum pro viribus violentissime / aciem longam ensis tui, capiti eius infligito, at si is ictus non fiat quam celerrime, partes posteriores tu feres.

C93, 228r

Der lanng spieß gögen dem schwert.

Item hallt dich also mit Disem stuckn wann Ir zü Samen gond / stand mit deinem lincken füoß vor unnd hallt dein spieß mit deiner lincken hand vornen gögen dem man dein rechte / hinder der deinem rechten schennckel Inn deinem spieß auf deiner rechten seiten Indes stich Im behend nach seinem geischt.

Sticht er dir also oben zü unnd dü mit deinem rechten fuoß gögen Im steest unnd helst dein schwert Inn dem Wechsel auf / deiner lincken seiten den Ort gögen der erden so gee aüf mit deiner schwert aüß dem Wechsel haw Im damit nach / seinem spieß so nimbstü Im seinen stich hinweckn aüf dein rechte seiten.

Hat Er dir also nach deinem spieß gehawen so / zückn deinner spieß unndersich Inn dem gangg wider aüf das dü deinem spieß wol vornen Im der mitthabest unnd stich / Im durch seinen leib seiner lincken seiten.

Hatt er dich also durch stochen durch deinen leib so haw Im mit sterckn mit deiner / lanngen schneid nach seinem haüpt.

The lance against the longsword.

Position yourself in the fight as follows: place your left foot forward, hold the lance with your left hand forward, toward the enemy, and your right hand behind your right foot, on the right side, and quickly thrust into his face.

If he attempts this same thrust to your face while you are leading with the right foot, holding your sword in *wechsel*[1] on the left side, its point tilted toward the ground, then raise it up and strike his lance onto your right side.

If he strikes your lance away thusly, then change under his sword, raise the lance up, strongly grip the middle of the shaft, and thrust through his left flank (as depicted by the left figure).

If he changes you through to thrust at your body, then strike your long edge into his head with all your might (as depicted by the right figure), but also note that if this strike does not happen as quickly as possible, you will suffer.

---

1. German, "changer." See Latin *habitus mutatorius*, "change guard."

Various 4

Cod. Vinob. 10826, 13v

Bipennis habitus contra hostem qui te cum ense pugnat.

Si habitu praedicto athletice uti volueris, dextrum pedem praepones, bipennem manu dextra iuxta inferiorem eius partem femori adplicatam / dextro contineas, sinistra vero medio contra hostem, mucro sursum vergat, inde mucronem anteriorem in visum / infigito.

Sin autem eadem ratione adversarius te adgreditor, in libra consistentem, nec non sinistrum praeposueris, ensis capulum manu / utraque in latere sinistro contineas, mucronem contra hostem porrectum, aciem ensis tuis longam eius bipennis adiungito, eoque / habitu impetum hostilem repellas velim. Verum si id perfeceris dextro introgreditor, atque ense ultra caput tuum sublato manibus / instar crucis formatis, caput ipsius hostis superne aciem longam praemittendo ferias.

Adversario autem eadem moliente, bipennem / retro caput collocatam anteversum exeras, atque vicissim lamina bipennis caput eius quassato.

C93, 228v

Die Hellenpart gögen dem schwert.

Item wann Fr zu samen gond so halst dich / also mit disem stuckn stannd mit deinem rechten fuoß vor unnd halst dein hellenparten mit deinem / rechten hannd Inn deinem unndern Ort aüf deinner rechten hüff dein lincke Inn der mit gögen / dem man den Ort ubersich Indes stich Im mit deinem vordern Ort zu seinnem gesicht.

Sticht er / dir also oben zü unnd dü aüch mit deinem lincken füoß gügem Im steest wol Inn der wag unnd helst dem / schwert mit baiden hennden Inn deinem hefft aüf deiner lincken Seiten den Org gögen dem man / so seß Im mit deinner lanngen schneidan sein hellenparten unnd seß Im das damit ab Indes trit mit / deinem rechten schennckel hinnein unnd gee aüf mit ... Armen mit deinem schwert ober / dein haüpt unnd ahw Im mit Lannger schneid oben nach seinem haüpt.

Hawt er dir also oben zü so / wind dein hellenparten hinder deinem haüpt herfür unnd haw Im aüch mit deinem plat deiner hellenparten / nach seinem haüpt.

(Another technique for) the halberd against the longsword.

Position yourself thusly for the aforementioned technique: place your right foot forward and hold the halberd with your right hand near the back point, on your hip, and your left hand in the middle, toward the enemy, with the front point tilted up; then, thrust your front point into his face.

If he thrusts at you in this same way while you are in a balanced stance with your left foot forward, holding your sword on the left side with both hands on your hilt and the point extended toward him, then set his halberd aside with your long edge, simultaneously step to the inside with your right foot, raise the sword above your head with crossed arms, and strike your long edge into his head (as depicted by the right figure).

If he strikes at you thusly, then wind the halberd behind your head and strike your blade into his head (as depicted by the left figure).

Various 11

Cod. Vinob. 10826, 19r

Venabuli habitus contra bipennem.

    Si congrediminis adinvicem, volesque inde ex palaestra palmam auffere, sinistrum praefigere memineris, venabulum in latere / dextro manu itidem dextra teneas, laeva sit medio venabulo adplicata contra hostem, atque ex ea forma corporis hostilis latus / sinistrum figito.

    Sed si idem is contra te moliatur, dextrum praeponentem, bipennemque iuxta partem eius posteriorem in latere / dextro manum dextra, laminam contra hostem tenentem, tum lamina tuae bipennis impetum hostilem in latus tuum sinistrum repellas, / at confestim insequutus laevo, pudenda adversariis punctione adpetito.

    Verum si idem hostis conabitur, sinistrum reducas / parte venabulis priori hostis impetum removebis in latus sinistrum.

    Sed si te eadem ratione repellere conabitur, laminam bipennis / tuae capiti eius infligito.

    Sin autem eveniat, ut is te superne eodem modo adgreditur, ictus hostili obviabis, atque pugione / supra brachium eius sinistrum gubernando collum hostis si transfixeris, concidere adversarius cogetur nescius vitae.

C93, 232r

Der schwein spieß gogen der hellenparten.

    Item hallt dich also mit Disem stuckn wann Ir zu samen gond / stannd mit deinem lincken füoß vor unnd halst dein speiß mit deiner rechten hand aüf deiner rechten seiten / dein lincke mitten Inn der stanngen gogen dem man Indes stich nach seinem leib seiner linckten seiten.

    Sticht / er dir also zu unnd dü mit deinnem rechten füoß gegen Im steest unnd helst dein hellenparten mit deinem rechten / hand bez deinem hindern Ort aüf deiner rechten seiten dein plat gogen dem man so nimb Im das ab mit deinem / plat aüf dein lincke seiten Indes volg mit deinem lincken schennckel hinnach unnd stich Im nach seinem gemechten.

    Sticht er dir also unnden zü so trit mit deinem lincken schennckel zü rückn unnd nimb Im das ab mit / deinem vordern tail deins speiß aüf dein lincke seiten.

    Hat Er dir das also abgenomben so haw Im mit deinem / plat deinner Hellenparten nach seinem haüpt.

    Hawt er dir also oben zü so ganng seinem haw entgogen unnd / stich Im ober seinem lincken arm durch seinem hals so stichstu Im zu der erden.

The boar spear against the halberd.

As you approach each other, position yourself as follows: place your left foot forward and hold the boar spear with your right hand on the right side and your left hand in the middle of the shaft, toward the enemy; then, thrust into the enemy's left flank.

If he thrusts at you in this way while you are leading with the right foot, holding the halberd on your right side with your right hand near the back point and the blade toward him, then set his attack onto your right side with your blade, immediately pass forward with your left foot, and thrust into his groin.

If he attempts this same thrust to your groin, then bring your left foot back and set his attack onto your left side with the front part of your boar spear.

If he parries you thusly with his boar spear, then strike your halberd's blade into his head (as depicted by the right figure).

If an unskilled adversary strikes you from above in this same way, then block his strike and thrust over his left arm through his throat, thus killing him.

Various 12

Cod. Vinob. 10826, 19v

Alia forma certaminis ex Venabulo contra Bipennem.

Quum athletice hoc habitu praedicto uti volueris contra hostem procedens, pedem sinistrum praeponas, iuxta mucronem inferiorem Venabulum / dextra tenere non dubites, medio sinistra sit adplicata, in alto coram tuo conspectu, caputque adversariis animose Venabulo / pulsabis.

Sed si is vicissim caput tuum concutere moliatur, te sinistrum itidem praefigentem, atque bipennis tuae mucronem posteriorem dextra / secundum caput tenente, sinistro vero adplicata eidem iuxta laminam contra hostem, tum lamina bipennis versus tuum latus dextrum / eius impetum removeas volo. Verum mucrone anteriori subito hostis pectus configito.

At si is te figere conabitur, in latus / dextrum repellito adversariis punctionem Venabulis impetu, inde vero dextro pede introgreditor, venabulo abiecto, manu / sinistra circa collum hostis regenda humerum dextrum adpraehende, dextra vero pedem eius sinistrum, et si ex eo habitu sursum / hostem levaris, pronum eum prosternes, prostratum itaque tuo arbitrio tractabis.

C93, 232v

Mer der schwein spieß gogen der hellenparten.

Item wann Fr zü samen Gond so hallt dich also / mit disem stuckn stannd mit deinem lincken füoß vor unnd hallt dein spieß mit dienner reichten hand / In deinem unndern Ort dein lincke Inn der mitt Inn der hoch vor deinem gesicht Indes haw Im nach / seinem haüpt.

Hawt er dir also oben zü unnd dü aüch mit deinem lincken füoß gögen Im steest und helst / dein hellenparten mit deinner rechten hand bey deinem hindern Ort hinder deinem haüpt dein lincke / In deiner stanngen bey dem plat gögen dem man so nimb Im das ab mit deinem plat aüf dein recht seiten / Indes stich Im mit deinem vordern Ort zü seiner prüst.

Sticht er dir also zü so nimb Im das ab mit deinem / spieß aüf dein rechte seiten Indes trit mit deinem rechten schennckel hinnein laß damit deinem spieß / fallen Inn dem greiff Im mit deiner lincken hand hinder seinem hals hinumb aüff sein recht achsel / unnd mit deiner rechten nach seinem lincken schennckel heb damit starckn obersich so wirfstu In aüff / das angesicht. Hastu In also gerworffen so magstü wol deinem spieß widerumb fallen unnd deines gefallens mit Im hanndlen mit hawen oder stichen.

Various Weapons

Another technique for the boar spear against the halberd.

To use this technique against an advancing enemy, position yourself as follows: place your left foot forward and hold your boar spear before your face in the high guard, with your right hand near the back point and your left hand in the middle (as depicted by the left figure); then, immediately strike his head.

If he strikes at your head in this way while you are leading with the left foot, holding the halberd with your right hand behind your head, by the back point, and your left hand on the shaft near the blade, toward the opponent (as depicted by the right figure), then set his attack onto your right side with your blade and simultaneously thrust your front point into his chest.

If he attempts this same thrust to your chest, then set him onto your right side with the boar spear, step to the inside with your right foot, drop your weapon, grasp his right shoulder by guiding your left hand around his neck, grab his left leg with your right hand, lift him up, and throw him onto his face. Then, retrieve your spear and deal with him as you like.

Various 22

Cod. Vinob. 10826, 120v

Habitus bipennis contra acinacem.

In congressu mutuo hac ratione te componito, sinistro pede contra hostem procedas, atque ictu superno caput ipsius saucies.

At si adversarius pari modo te adgreditur superne dextrum praeponentem, tum sinistro subsequutus, in habitum coronae acinacem / tollas, et hostilem ictum excipias.

Sin autem is tuum impetum eadem ratione averterit, tum dextro versus ipsum pede progreditor, / nec non inferiorem mucronem ab inferne contra hostis visum vel pectus torqueas.

Verum si is mucronem consimilis / modo contra te torserit, tunc ex coronae habitu acinacem, bipennis eius superiniscias, supprimasque, et eo pacto vis hostilis est / repulsa. Inde autem mucronem posteriorem manu sinistra corripias adversariis bipennis, eademque itaque bipennem firmiter / contineas, post subsequutus dextro, celeriter caput eius saucies.

Sed si is eum impetum exceperit, tum pectus ipsius vel / visum fodito, atque post ictu gemino recedas.

Cod. Vinob. 10826, 120v

Ein stuck in der hellenbarten gegen dem Duseggen.

Item schick dich also in dises stuck mit dem züfechten, trit mit deinem linggen schenckel hinein, und / . . . .

. . . also oben aüf dich, und dü mit deinem rechten / füoß vorsteest, so trit mit deinem linggen schennckel hinein, und haw mit deinem Duseggen wol / aüff in der kron fur dein haübt. verstey Im damit seinen häw.

Hat er dü deinem häw also / abgenomen, so trit mit deinem rechten schenckel hinein, und wind Im deinen vordern ort von / under aüff zü seinem gesicht oder prüst.

Hat er dir also mit seinem ort . . . / so fall Im mit der kron aüff sein hellenbarten, und truckn damit wol wider, so nimbst dü / Im das ab, in dem greiff Im mit deiner linggen hand in seiner hindern ort, faß damit sein / hellenbarten, und trit mit deinem rechten schenckel hinein, und häw Im indes nach seinen / haupt.

Verstez er dir dann das, so stich Im nach seinem gesicht oder der prüst, Indes / zeuch dich mit ainem Dopplehäw von Im Ruckn.

The halberd against the dussack.

Position yourself for this technique as follows: step toward the enemy with your left foot and strike an *oberhaw*[1] to his head (as depicted by the right figure).

If he strikes an *oberhaw* to your head in this way while your right foot is leading, then step to the inside with your left foot and displace his attack by striking your dussack up into *kron*[2] (as depicted by the left figure).

If he displaces your *oberhaw* thusly, then step to the inside with your right foot and wind your back point[3] into his face or chest from below.

If he winds his back point toward you from below, then displace him by striking your dussack from *kron* and pressing down on his halberd. At the same time, seize his back point with your left hand, step to the inside with your right foot while gripping his halberd firmly, and strike his head. If he parries this attack, then thrust into his face or chest and withdraw from him with a *dopplehaw*.[4]

---

1. German, "over strike; strike from the upper openings." See Latin *ictus superus*, "above strike."
2. German, "crown." See Latin *habitus coronae*, "position of the crown."
3. The German text reads "forward point."
4. German, "double strike." See Latin *ictus geminus*, "twin strike."

Various 23

Cod. Vinob. 10826, 121r

Alius praecedentis habitus.

In hoc certamine sinistrum pedem praeponas, atque bipenne thoracem hostis pungas.

Verum si is pari modo te pungere conetur / sinistrum praeponentem, tunc in habitu coronae acinacem subleves, atque punctionem hostilem repellas, inde bipennem ipsius / mediam, manu sinistra adpraehendas, et eo modo sinistrorsum deprimas, nec non mucronem acinacis tuis visus hostis impingas.

Verum si is pari ratione te fuerit adgressus, tum intra utramque manum bipenne excipias sinistrorsum.

Quod si idem is exceperit, / tu acie longa premissa caput adversariis saucies, atque gemino passu pungendo et feriundo recedas.

Cod. Vinob. 10826, 121r

Ein anders stuck in der hellenbarten gegen dem Duseggen.

Item halt dich also mit dem zufechten in disem stuck, Trit mit deinem linggen schenckel hinein, / und stich Im deiner hellenbarten nach seinem leib.

Sticht er also aüff dich, und dü mitt / deinem linggen füoß gegen Im steest, so haw aüff mit deinem Duseggen, in der kron, nimb Im / damit seinem stich hinwegt, Indes greif mit deinem linggen hand mitten in sein hellenbarte / truckn damit wol undersich aüf dein linggen seiten gegen der erden, Indes stich Im mit deinem / ort deines Duseggen zü seinem gesicht.

Sticht er also aüf dich, so nimb Im das ab zwischen / baiden henden in dein hellenbarten aüff dein linggen seiten.

Hat er dir das also abgenomen, / so haw Im mit deinem Düseggen mit langer schneid zü seinem haübts, und wind dich mit stiche / unnd hawen zwifach von Im.

Another technique for the halberd against the dussack.

In this fight, step forward with your left foot and thrust your halberd into the enemy's chest.

If he thrusts his halberd at your chest while your left foot is leading, then parry him by striking your dussack up into *kron*[1]; at the same time, seize the middle of his halberd with your left hand, push it down onto your left side, and thrust the point of your dussack into his eyes (as depicted by the left figure).

If he thrusts his dussack at your face in this way, then catch him between your hands with the halberd and set him onto your left side (as depicted by the right figure).

If he parries you thusly, then strike your long edge into his head, and double-step away from him while thrusting and striking.

---

1. German, "crown." See Latin *habitus coronae*, "position of the crown."

Various 24

Cod. Vinob. 10826, 121v

Habitus ex ense hispano contra venabulum.

In mutuo congressu hoc modo te adaptes, sinistrum pedem contra hostem colloces, atque venabulo hostis visum vel pectus / pungas.

Veru si is eodem habitu utatur, dextrum pedem te praeponente ensem hispanum continens, tunc obviam ferias, / ipsius punctionis, atque iuxta cuspidem venabulis eum adplices. Inde si sinistrorsum represseris, punctionem eius / avertisti, atque post venabulum de parte priori corripias.

Quod si is tuum corripere conabitur. Venabulo retracto, / de hostis latere sinistro versus dextrum, pectus pungito seu visum.

Sin vero adversarius pari modo idest gemina / punctione te fuerit adgressus, tum ense hispano dextrorsum repellas, postea sinistro pede intro concedas, mediamque / hastam ipsius adpraehendas, et celeriter visum ipsius fodias. Sin autem id removerit, tum ab ipsius hasta transmutabis, / inde si in triangulum prosilieris caput hostis ferias.

Cod. Vinob. 10826, 121v

Ein stuck in dem rapier gegen dem schwein spies.

Item schich dich also in dises stuck mit dem zufechten, Trit mit deinem linggen schenckel zü Im hinein, / und stich Im mit deinem schweinspies zü seinem gesicht oder der prüst.

Sticht er also aüff dich, / und dü mit deinem rechten füoß gogen Im steest mit deinem Rapire, so haw seinem stich entgegen / setz Im damit vornen an seinen spieß, Indes truckn von die aüf dein linggen seiten, so nimbst dü / Im seiner stich hinweckn, greiff Im damit vornen nach seinem speiß.

Will er dir deinem speiß / also ergreiften, so zuckn behend an dich, und stich Im von seiner linggen aüf sein rehct seitten / zü seiner prüst oder dem gesicht.

Sticht er also zwifach aüf dich, so nimb Im das ab aüff / der rechte seitten mit deinem Rapire, In dem trit mit deinem linggen schenckel hinein, unnd / greiff mit deiner linggen hand mitten in seiner Speiß, und stich Im indes behend zü seinem / gesicht, versety er dir das, so wechsel durch an seinem speiß, und spring in triangel, / unnd häw Im nach seinem haubts.

The rapier against the boar spear.

Position yourself for this technique as follows: step toward the enemy with your left foot and thrust your boar spear into his face or chest.

If he thrusts at you in this way while you are holding the rapier with your right foot forward, then strike his weapon near the spearhead (as depicted by the left figure), displace his thrust onto your left side, and seize the forward part of his boar spear.

If he tries to seize your boar spear, then quickly pull it back and thrust into his face or chest, from his left to his right side.

If he thrusts twice at you with this same technique, then set him onto your right side with the rapier, step to the inside with your left foot, and seize the middle of his spear with your left hand while simultaneously thrusting into his eyes. If he parries, then change through on his spear, spring forward in a triangle, and strike his head.

Appendix A

Latin Glossary

CONCISE GLOSSARY OF LATIN TERMS

abl.   ablative case
acc.   accusative case
adj.   adjective
adv.   adverb
conj.   conjunction
coord.   coordinate
dat.   dative case
dep.   deponent
f. feminine gender
impers. impersonal verb
irreg.   irregular usage
lit. literally
m. masculine gender
n. neuter gender
prep.   preposition
w. with

A

a *or* **ab**, *prep.* + *abl.*, from, away from
ab invicem, *idiom,* from each other; *irreg.,* toward each other
abeo, -ire, -ivi, -itum, withdraw, go away; *w. weapons,* drop, release, get rid of
abiecto, -are, -avi, -atum, throw down
abigo, -ere, abigi, abactum, repel, parry
absum, -esse, -fui, -futurum, be free, freed from
abtango, -ere, abtetigi, abtactum, touch, strike away

abtingo, *see* **abtango**
ac, *conj.,* and, and also
accomodo, -are, -avi, -atum, adjust, apply to, adapt to
acies, aciei, *f.,* sharp edge of a blade
acies brevis, *f.,* short edge, false edge
acies longus, *f.,* long edge, true edge
ad, *prep.* + *acc.*, toward, to, into
adeo, *adv.,* indeed
adgredeo, -ere, adgressi, adgressum, attack, advance, approach
adgredior, adgredi, adgressus sum, *dep., see* **adgredeo**
adhibeo, -ere, adhibui, adhibitum, apply, include, add
adjungo, -ere, -adjunxi, -adjunctum, add, attach, apply to, harness
admoveo, -ere, admovi, admotum, bring to, apply
adpareo, -ere, -ui, -itum, *corrupted form of* **appareo,** be evident, visible
adpeto, -ere, -ivi, -itum, aim for, attack
adplico, -are, -avi, -atum, connect, bring into contact with, press, apply; *w. weapons,* to hook
adpraehendo, -ere, adpraehendi, adpraehensum, seize, grasp, take hold of
adversus, *adv. and prep.* + *acc.,* against, opposite, facing
affligo, -ere, afflixi, afflictum, throw down, injure, crush
animadverto, -ere, animadverti, animadversum, notice, anticipate
animose, *adv.,* energetically, in an animated manner
antepono, -ere, anteposui, antepositum, *w. dat.,* place in front of
appeteo, -ere, appetivi, appetitum, *see* **adpetere**

apporto, -are, -avi, -atum, carry, transport
arripio, -ere, -ui, arreptum, seize, snatch, take hold of, pull a thing to one's self
ars athletica, -ae, *f.*, athletics
athletice, *adv.*, athletically
attollo, -ere, *see* **tollo**
aut, *conj.*, or (**aut . . . aut,** either . . . or)
autem, *conj.*, on the other hand, however, moreover, also
averto, avertere, averti, aversum, avert, parry, set aside (*see German* **absetzen**)
axillus, -i, *m.*, *irreg.*, armpit; *diminutive form of* **axis, -is,** *m.*, axis

B

bipennis, -is, *f.*, halberd, double-edged battle axe (*see German* **hellenbarten**)

C

caeterum, *see* **ceterum**
cancellatus, -a, -um, *adj.*, crossed
capulus, -i, *m.*, hilt, handle, shaft
castrum, castri, *n.*, guard, defense; *lit.* fortress
certamen, certaminis, *n.*, fight, struggle, contest
cesso, -are, -avi, -atum, cease, be free of
ceterum, *adv. and conj.*, but, however
circumago, -ere, circumegi, circumactum, turn around
circumflecto, -ere, -flexi, -flexum, *w. weapons*, wind (around), change through (*see German* **wechseln**)
circumspecte, *adv.*, with circumspection, cautiously
circumvoluto, -are, -avi, -atum, *see* **circumflecto**
circumvolvo, -ere, -volvi, -volutum, *see* **circumflecto**
coarto, -are, -avi, -atum, move in, close; confine, press together
collum, colli, *n.*, throat, neck
commode, *adv.*, at the right moment
compono, -ere, composui, compositus, to compose oneself, position one's body
conatus, -us, *n.*, thwarted attack, attempt
concutio, -ere, concussi, concussus, strike violently
conficio, -ere, confeci, confectum, finish off, defeat finally
conjicio, -ere, conjeci, conjectus, thrust, throw
collisio, -nis, *f.*, clash, bind
colloco, -are, -avi, -atum, position, place
comprehendo, -ere, comprehensi, comprehensus, catch, grasp; *w. weapons*, hook
concedo, -ere, concessi, concessus, *irreg.*, take a passing step (forward); *lit.*, concede, depart
confestim - immediately, at once (*see German* **indes**)
congressus, -us, *m.*, approach, fight, bind (*see German* **zufechten**)
conor, conari, conatus sum, *dep.*, try, attempt
conquasso, -are, -avi, -atum, strike violently, break, smash
conquiro, -ere, conquisivi, conquisitum, probe, seek out, search for
consequor, consequi, consecutus sum, *dep.*, take a passing step (forward), *lit.*, follow, pursue
consimilis, -is, *adj.*, similar
consisto, -ere, constiti, constitum, to stand, pause, halt, linger, take a position
contactus, -us, bind
contineo, -ere, continui, contentum, hold, grip
contingo, -ere, contigi, contactum, come into contact with; *w. weapons*, bind
contorqueo, -ere, contorsi, contortum, wind, thrust
contra, *adv., and prep. + acc.*, against, toward, opposite, facing
converso, -are. -avi, -atum, turn (around), wind
cubitus, -us, *m.*, elbow
cura, -ae, *f.*, care, attention
custodia, -ae, *f.*, guard
custodia aperta, *f.*, open guard (*see German* **offenhut**)
custodia cancellata, *f.*, crossed guard (*see German* **schrankhut**)

D

de, *prep. + abl.*, from, out from, away from, out of; *irreg.*, on
de improviso, *idiom*, unexpectedly, suddenly, without warning
declino, -are, -avi, -atum, decline, lower, tilt downward
defensio, -nis, *f.*, defense, guard
deflecto, -are, -avi, -atum, deflect, turn (one's body), change directions
deorsum, *adv.*, downward
derepente, *adv.* suddenly
diligenter, *adv.*, carefully
dirigo, -ere, direxi, directum, direct, aim, point
discedo, -ere, discessi, discessum, depart, withdraw
dolor, doloris, *m.*, pain, suffering
dorsum, dorsi, *n.*, back
dubito, -are, -avi, -atum, doubt, hesitate
dupliciter, *adv.*, twice

E

efficio, -ere, effeci, effectum, execute, bring about, cause, accomplish
effigies, -iei, *f.*, effigy, image
eludo, -ere, elusi, elusum, elude, parry, void
evenio, -ire, eveni, eventum, happen, turn out
ex, *prep. + abl.*, from, after
ex totis viribus, *idiom*, with all of one's strength
excipio, -ere, excepim exceptum, parry, ward off
excutio, -ere, excussi, excussum, cast off; search
exero, -ere, exerui, exertum, thrust out, extend
exter, extera, -um *adj.*, outer, on the outside

F

fallo, -ere, fefelli, falsum, deceive, trick; elude, escape; cause to fall, trip
femur, femoris, *n.*, thigh
ferio, -ire, hit, strike
figo, -ere, fixi, fixum, thrust, pierce
firmiter, *adv.*, firmly, tightly, securely
flecto, -ere, flexi, flexum, curve, wind, bend
fodio, -ere, -i, fossum, stab
formo, -are, -avi, -atum, form, create
fortiter, *adv.*, strongly, forcefully, with power

G

geminus, -a, -um, *adj.*, double
genu, -us, *n.*, knee
guberno, -are, -avi, -atum, wield, guide, direct (one's weapon)

H

habitus, -us, *m.*, technique, position, guard
habitus aversionis, *m.*, general term for a parry with a counterattack (*see German* **abnemen**); *lit.* method of setting aside
habitus liberi, m., free stance (*see German* **freystand**)
habitus mutatorius, *m.*, change-through, change guard (*see German* **wechsel**)
hasta, -ae, *f.*, staff, spear, pike, lance
hastile, -is, *n.*, shaft of a polearm
hastula, -ae, *f.*, shortstaff
haurio, -ire, hausi, haustus, *irreg.*, thrust; *lit.*, draw up/out

humerus, -i, *m.*, shoulder
humus, -i, *f.*, ground, earth

I

ic(i)o, icere, ici, ictum, strike, stab
ictus, -us, *m.*, strike
ictus inferus, *m.*, strike from the lower openings (*see German* **unterhau**)
ictus superus, *m.*, strike from the upper openings; *lit.* above strike (*see Geman* **oberhau**)
idest, *adv.* (*contraction*, **id** + **est**), i.e., that is, specifically, in other words
impedio, -ire, -ivi, -itum, impede, block, hinder, obstruct
impeditus, -a, -um, *adj.*, impeded, blocked, hindered, obstructed
impello, -ere, impuli, impulsum, thrust, impale
impeto, -ere, impetivi, impetitum, *see* **peto**
impetus, -us, *m.*, attack; impetuousness
impingo, -ere, impegi, impactum, *see* **impello**
impulso, -are, -avi, -atum, *see* **impello**
impulsus, -us, *m.*, thrust
in, *prep. + abl.*, in, on, during
in congressu, *idiom*, during the approach (*see German* **zufechten**)
in primis, *idiom*, above all, particularly, especially
incisio, -nis, *f.*, incision, slice, cut (*see German* **schnit**)
incurro, -ere, incurri, incursum, rush in, run toward (*see German* **einlauffen**)
incursio, -nis, *f.*, rush-in, attack, incursion
incurso, -are, -avi, -atum, *see* **incurro**
indefensio, -nis, *f.*, opening
indefensus, -a, -um, *adj.*, undefended, exposed, open to attack
inferus, -a, -um, *adj.*, low, underneath
infigo, -ere, infixi, infixum, fix, fasten (on); thrust into
infirmiter, *adv.*, weakly
inflecto, -ere, inflexi, inflexum, *see* **flecto**
inflexio, -onis, *f.*, curve, wind, bend
infligo, -ere, inflixi, inflictum, strike or thrust into
inicio, -ere, injeci, injectum, strike into
iniungo, -ere, iniunxi, iniunctum, unite
insequor, insequi, insecutus sum, *dep.*, take a passing step to the inside, *lit.*, follow, pursue
inserto, -are, -avi, -atus, thrust into
instantia, -ae, approach
interclusio, -nis, block, technique that immobilizes the opponent's weapon

interea, *adv.*, simultaneously, at the same time, meanwhile (*see German* **indes**)
interim, *adv.* at that moment, meanwhile
intorquo, -ere, intorsi, intortum, turn around, rotate, wind, shoot in
intorsio, -nis, *irreg., from* **intorqueo**, turning motion, winding, shooting-in
intra, *adv.*, inside, within, under
intrinsecus, *adv.*, on the inside, from within, inward
intro, *adv., see* **intra**
introgredior, introgredi, introgressus sum, *dep.*, enter, step to the inside
invicem, *adv.*, in turn, mutually
irritus, -a, -um, *adj.*, ineffective, useless, in vain
irruptio, -nis, *f.*, attack, entry
ita, *adv.*, thus, so, in such a way that
itaque, *adv.*, thus, consequently
itidem, *adv.*, thusly, in the same manner, likewise, similarly, also
iuxta, *prep. + acc.*, near, close to

L

laedo, -ere, laesi, laesus, strike, wound
lancea, -ae, *f.*, lance, longstaff, pike
letalis, -is, -e, *adj.*, lethal, deadly

M

manus, -us, *f.*, hand
memini, meminisse, remember
mola, -ae, *irreg.*, knee; *lit.* jawbone
molior, moliri, molitus sum, *dep.*, attempt
mucro, -nis, *m.*, point of a weapon (*see German* **ort**)
mucro anterior, *m.*, front point, tip (*see German* **vordern ort**)
mucro brevis, *m.*, short point, butt
mucro longus, *m.*, long point, tip
mucro posterior, *m.*, back point, butt (*see German* **hindern ort**)
mucro recurvus, *m.*, spike of a halberd
mutatorius, -i, *m.*, change strike (*see German* **wechselhaw**)
mutatorius cancellatus, *m.*, crossed change strike
mutatorius simplicis, *m.*, simple change strike
muto, mutare, -avi, -atum, change, shift, switch (*see German* **wechseln**)

N

nec + non, *conj.*, and certainly, besides, be sure to
negotium, -i, *n.*, trouble, problem, activity
noceo, -ere, nocui, nocitum, *w. dat.*, harm, injure
num, *adv.*, now

O

oblido, -ere, oblisi, oblisum, crush
obliuiscor, obliuisci, oblitus sum, *dep.*, forget
observo, -are, -avi, -atum, observe, watch out for, anticipate; *in the bind*, feel
obsum, obesse, obfui, obfuturum, hurt
oportet, -ere, oportit, *impers.*, it is necessary, ought

P

palma, -ae, *f.*, prize
pariter, *adv.*, equally, together, also
pars, partis, *f.*, part, region, side
pectus, pectoris, *n.*, pectoral, chest (*see German* **prust**)
per, *prep. + acc.*, through, across; during; by means of
percussio, -nis, *f.*, strike
persequor, persequi, persecutus sum, *dep.*, follow, pursue; overtake, attack
pes, pedis, *m.*, foot
peto, -ere, petivi, petitus, attack, aim for
plaga, -ae, *f.*, strike
pono, -ere, posui, positum, position, place
poples, poplitis, *m.*, knee
porrigo, -ere, porrexi, porrectum, extend
porro, *adv.*, hereafter; again; on the other hand; in turn
postpono, -ere, -posui, -positum, *w. acc.*, place behind
praedico, -are, -avi, -atum, mention previously
praefigo, -ere, -fixi, -fictum, set in front, lead
praefixus, -a, -um, set in front, leading; *irreg.*, pointed, aimed
praenomino, -are, -avi, -atum, mention previously
praepono, -ere, -posui, -positum, position in front; *w. dat.*, place in front of
praescribo, -ere, -scripsi, -scriptum, mention previously
primo, *adv.*, first, in the beginning
primus, -a, -um, *adj.*, first, foremost, primary
pro viribus, *idiom*, with all of one's strength
procedo, -ere, processi, processum, proceed, advance
profligo, -are, -avi, -atum, throw to the ground

propello, -ere, propuli, propulsum, propel, drive forward, thrust
propugnaculum, -i, *n.,* defense
propulso, -are, -avi, -atum, parry
prosequor, prosequi, prosecutus sum, *dep.,* take a passing step (forward), *lit.,* follow, pursue
prosterno, -ere, prostravi, prostratum, prostrate, throw to the ground
protrudo, -ere, protrusi, protrusum, push forward, thrust
proverto, -ere, proverti, proversum, turn forward
pubes, -is, *f.,* groin
pudenda, -ae, *f.,* groin
pulso, -are, -avi, -atum, strike, attack
punctio, -nis, *f.,* puncture, stab, thrust
pungo, -ere, pupugi, punctum, puncture, stab, thrust

Q

quacumque, *adv.,* wherever
quasi, *adv. and conj.,* as if
quasso, -are, -avi, -atum, strike, batter, flourish
quoque, *adv.,* too
quum, *adv.,* when, while, as, since, although; *prep. + abl.,* with, at the same time as

R

recedo, -ere, recessi, recessum, withdraw
recipio, -ire, recepi, receptum, take back, bring back, draw back
recta, *adv.,* right, straight, directly
recte, *adv.,* properly, correctly
recurvus, -a, -um, *adj.,* curved
reddo, -ere, -idi, -itum, render, pay back
reduco, -ere, reduxi, reductum, bring back
refero, -ere, rettuli, relatum, bring back
refringo, -ere, refregi, refractum, check, put an end to, shatter, destroy
rego, -ere, rexi, rectum, *w. weapons,* wield; *lit.* guide, direct
rejicio, -ere, rejeci, rejectum, drive back, displace
reiscio, *corrupted form of* **rejicio**
repello, -ere, repulli, repulsum, repulse, fend off, push back, parry
repente, *adv.,* suddenly
reperio, -ire, repperi, repertum, discover, learn; *in the bind,* feel

respicio, -ere, respexi, respectum, look at, face; *w. weapons + acc.,* point toward
resto, -are, -avi, -atum, stand one's ground, resist
retorqueo, -ere, retorsi, retortum, turn back, rotate backward, wind
retraho, -ere, retraxi, retractum, draw back, withdraw, bring back
retrorsum, *adv.,* backward, behind
rite, *adv.,* rightly, in the usual way, properly
roto, -are, -avi, atum, rotate, wind around
rursus, *adv.,* back, backward; on the other hand; again, a second time

S

saucio, -are, -avi, -atum, wound, hurt, stab
securis, -is, *f.,* poleaxe
securis letalis, *f.,* murder-axe (*see German* **mordagst**)
sedulo, *adv.,* carefully
seu, *conj.,* or
si, *conj.,* if
si vero, *idiom (expresses skepticism),* if somehow, if in fact, if really
solvo, -ere, solvi, solutum, release, unbind
statim, *adv.,* at once
statuo, -ere, statui, statutum, stand, position, place
status, -us, *m.,* stance, position
status liborum, *m., corruption of* **status librarum,** balanced stance (*see German* **die wag**)
stimulus, -i, *f.,* spike of a halberd
strenue, *adv.,* with strength, vigorously, forcefully
sub, *prep. + abl.* under, beneath; *w. acc.,* up to, toward, into, near
sub ictum, *idiom,* within range
subito, *adv.,* quickly, immediately, at once
superne, *adv.,* from above, from the upper openings
supernus, -a, -um, *adj.,* above, upper, high
supinus, -a, -um, *adj.,* supine, flat on one's back, lying face upward
supra, *adv.,* above; *prep. + acc.,* above, over
sursum, *adv.,* up, on high

T

tango, -ere, tetigi, tactum, touch; *w. weapons,* bind
teneo, tenere, tenui, tentus, hold
tollo, tollere, sustuli, sublatum, raise up, lift
torqueo, -ere, torsi, tortum, turn, wind

transitio, -nis, *f.*, passing-over, over-going (*see German* **obergehen**)
transmitto, -ere, transmisi, transmissus, send across, change through
tundo, -ere, tutudi, tusum, strike, beat, crush
tute, *adv.*, safely

U

unde, *adv.*, from what, from where, from which
undique, *adv.*, on both sides, everywhere
usurpo, -are, -avi, -atum, usurp, use against you
ut, *conj.*, so that, as
uterque, utramque, utrumque, *adj.*, both (of two); *adv.*, on both sides
utor, uti, usus sum, *dep.*, use, employ

V

vel, *conj.*, or; **vel . . . vel,** *coord.*, either . . . or
verbo,

vero, *adv.*, certainly, definitely, actually; *irreg. (as conj.)*, and, also
versus, *prep. + acc.*, toward, into
verum, *adv.*, however, yet, in fact, truly, indeed; *irreg. (as conj.)*, and, also
viadverto, -ere, viadverti, viadversum, turn aside, parry
vibro, -are, -avi, -atum, propel, shoot, thrust; wave, flourish
vicissim, *adv.*, in turn, again
vis, vis, *f.*, hostile force, violence, strength, might, power; *irreg.*, attack
visus, -us, *m.*, vision, sight, eyes; *irreg.*, face (*see German* **gesicht**)
volo, velle, volui, wish, want
volvo, -ere, volvi, volutum, wind
vultus, -us, *m.*, face (*see German* **gesicht**)

APPENDIX B

GERMAN GLOSSARY

CONCISE GLOSSARY OF EARLY NEW HIGH GERMAN TERMS

acc. accusative case
adj. adjective
adv. adverb
conj. conjunction
coord. coordinate
dat. dative case
impers. impersonal verb
irreg. irregular usage
lit. literally
nom. nominative case
prep. preposition
v. verb
w. with

A

ab, *prep.*, aside, off
abgewinnen, *v.*, win, be victorious
abnemen, parry, general method of setting-aside (*see* Latin **habitus aversionis**); *lit.*, taking off
absessen, *see* **absetzen**
absetzen, *v.*, turn away, avert, set aside (*see* Latin **averto**)
abwechseln, change; *w. weapons*, act of changing an attack or guard
abwinden, *v.*, wind off; *w. weapons*, wind away a thrust

alber, low guard; *lit.*, fool
an, *prep.*, on
anbinden, *v.*, bind on, press one's weapon upon the opponent's weapon
anbunden, *see* **anbinden**
angesicht, *see* **gesicht**
arm, arm
aschel, shoulder
arbait, work, task (of defeating one's opponent)
ansetzen, *v.*, set on; *w. weapons*, make contact with the opponent's body (by either a thrust or strike, though usually the former) (*see Latin* **contactus**)

B / P

bainbruch, throw achieved by targeting the opponent's legs; *lit.*, leg break
behend, *adv.*, quickly
bey, *prep.*, by, near
bind, bind, moment of actual contact between two weapons
binden, *v.*, bind with the opponent
pflug, *w. weapons*, middle guard; *lit.* plow
blat, blade
bloß, *adj.*, undefended, unarmored, open
brust, breast, chest, pectoral (*see Latin* **pectus**)
bund, *see* **bind**

183

D / T

damit, *adv.*, with this
darauß, *prep.*, from
dein, *adj.*, your
drucken, *v.*, press, push down
dü, *nom. pron.*, you
durch, *prep.*, through
durch . . . wechsel, *idiom*, change line of attack, move the point to another opening
dussegen, dussack
dich, *acc. pron.*, you
trianngel, triangle
trit, step
trit hinein, inside step
trit ruckn, back step
trit inn trianngel, triangle step
trucken, *v.*, press, push

E

einbrechen, *v.*, break in (*see Latin* **irruptio**)
einhawen, *v.*, strike on the inside
einlauffen, *v.*, charge forward, rush in (*see Latin* **incurso**)
einsheissen, *v.*, thrust at the opponent while maintaining pressure on his weapon; *lit.,* shoot in
einwinden, *v.*, wind in
ellenbogen, elbow
entgogen, *prep.*, together; against
er, *nom. pron.*, he

F / V

fallen, *v.*, fall on, down upon
verkert, *adj.*, inverted
verkerter leger, inverted guard
verkerter schlag, inverted strike
versaßung, defense
verseßen, *see* **versetzen**
versetzen, *v.*, displace
vier, four
vier blossen, four openings, four quarters
volgen, *v.*, follow
volgen . . . hinnach, follow from behind, pass forward
vom tag, high guard; *lit.* from the roof
von, *prep.*, of, from
vor, before; *w. weapons*, when one acts before the opponent
vorder, *adj.*, forward (*see Latin* **anterior**)
freylang, technique by which the weapon is released with one hand; lit., free long
freystand, open stance or guard; *lit.*, free stance (*see Latin* **habitus liberi**)
fuhlen, *v.*, feel; *w. weapons*, determine the opponent's strength at the bind
fuoß, foot
furbaß, *adv.*, onward
furgeseßen, *v.*, set forward, place in front
furgesetzen, *see* **furgeseßten**

G / K

geen, go
gegen, *prep.*, toward
gemecht, privates, groin (*see Latin* **pudenda**)
gerworffen, *v.*, throw
geschrenk, *adj.*, crossed
geschrenkt hut, crossed guard
geschrenkten wechsel, *lit.*, crossed changer (*see Latin* **mutatorius cancellatus**)
gesicht, face (*see Latin* **visus/vultus**)
gewalt, *adv.*, forceful, powerful
gleich, *adj.*, the same, identical, equal
gleichen fuoß, *idiom*, stance whereby feet are placed equally forward
gogen, *see* **gegen**
kopff, *m.*, head
krad, *irreg.*, strike
greiffen, *v.*, seize, grab, grip
kreuzweisen, *prep.*, crosswise
kron, crown; *w. weapons*, a secondary guard
krum, *adj.*, crooked, bent
krumb überfallen, crooked attack
krumphaw, crooked strike
kurß, *adj.*, short
kurtz, *see* **kurß**
kurtzen ort, short point
kurtzen schneid, short edge
kurtzhau, short cut or strike
gut, *adj.*, good

H

hacken, hook; action whereby one's leg hooks the opponent's leg
hacknen, heel kick

hacten, *see* **hacken**
halb, *adj.*, half
halben hallenbarten, method of holding the halberd with both hands spaced equally from the center; *lit.,* half-halberd
halben speiß, method of holding the spear with both hands spaced equally from the center; *lit.,* half-spear
halber stanggen, method of holding the staff with both hands spaced equally from the center; *lit.,* half-staff
hals, throat
halten, *v.*, hold
hand, hand
hart, *adj.*, hard; *w. weapons*, strong at the bind (*see also* **fuhlen**)
haubt, head
haw, cut, strike, hew
hellenbarten, halberd (*see Latin* **bipennis**)
hert, *see* **hart**
hinder, *adj.* behind, back (*see Latin* **posterior**)
hinein, *prep.*, in, inside
hinnach, *prep.*, from behind
hinter, *see* **hinder**
hinweckh, *prep.*, away
hoch, *adj.*, high; *w. weapons*, high guard (*see* **vom tag**)
hufft, hip
hut, guard (*see Latin* **custodia**)

I / J

im, *acc. pron.*, him
indes, *adv.*, simultaneously, at the same time, during (*see Latin* **confestim/interim**)
inwendiger, *prep.*, inside

L

leger, position, guard (*see German* **hut**, *Latin* **custodia**)
lang, long
langgen ort, long point; thrust with extended arms
lanngen schneid, long edge
langger sneide, *see* **lanngen schneid**
langger spiess, long spear
laß, loose, release, let
leib, body
lincken, left
linggen *see* **lincken**
lincken seiten, left side
loß, *see* **laß**

M

man, man, opponent, enemy (*see Latin* **adversarius/hostis**)
mit, *prep.*, with
mittle, middle
mittlehaw, middle-strike; horizontal strike from either the left or the right
mord, murder
mordagst, pole axe (*see Latin* **securis/securis letalis**); *lit.,* murder axe
mordhaw, murder strike

N

nach, *prep.*, after, toward
nachraise, *lit.,* after-strike
nachreise, *see* **nachraise**
nachtstoß, *lit.,* after-thrust
nemen, *v.*, take, remove
nimben, *see* **nemen**
nimb . . . ab, *idiom*, taking aside, parry, displacement (*see* **abnemen**)
nit, *adv.*, not

O

offen, *adj.*, open
offen hut, open guard (*see Latin* **custodia aperta**)
ober, *prep.*, upper, over
oberhaw, upper strike; any descending cut or strike from above the waist (*see Latin* **ictus superus**)
ob, *see* **ober**
ort, point of the weapon (*see Latin* **mucro**)
obergehen, *v.*, going/passing over (*see Latin* **transitio**)
orth, *see* **ort**
ochs, *w. weapons*, hanging guard; *lit.,* ox

R

raißen, *v.*, attack, wrench, force, pull
rappier, rapier
recht, *adj.*, right
rechten seiten, right side
reißen, *see* **raißen**
ruckh, *adj.*, back, behind

S

sturzhaw, plunge strike
schrankhut, barrier guard
stoß, thrust
stoßen, *v.*, thrust, drive, push
stich, thrust, stab
stichen, *v.*, thrust, stab
schlag, strike, blow
schlagen, *v.*, strike
speiß, spear
stanngen, staff
seßen, *v.*, set
seß … ab, *idiom*, set aside (*see* **absetzen**)
streichen, *v.*, strike a cut or blow
schnit, slice, cut
sterckh, *w. weapons*, portion from the hilt to the middle; *lit.* strong
stannd, stand
schwech, *adj.*, weak, (*see* **weik**)
schwechen, *v.*, weaken
schwert, sword
swert, *see* **schwert**
sweinspieß, boar spear.
schefftlin, an older, lesser known name for "javelin"
stuck, piece, technique
schickh, position, place
schneide, cutting edge
schennckel, ankle, leg
stehen, *v.*, stand
seiten, side
schleck, strike, blow
schleg, strike, blow
sperzen, block
sein, his
springen, *v.*, spring, leap
schneid, edge
seiten stoß, side thrust
seiten haw, side strike
scheißen, *v.*, shoot, thrust

U

überfall, attack
überge[h]en, *v.*, pass over, go over, transition, change
überwinden, *v.*, **wind over**
underhaw, lower strike; any ascending cut or strike that starts from below the waist (*see Latin* **ictus infernus**)
undern, *adj.*, lower, under, below
undern hut, lower guard
unterhaw, *see* **unterhaw**

W

wa[a]g, balanced stance (*see Latin* **status liborum**); *lit.*, scales
waik, *adj.*, weak; *w. weapons*, weak at the bind (*see also* **fuhlen**)
wechsel, change, changing guard (*see Latin* **circumflecto, mutatorium**)
wechsel … ab, act of changing an attack or guard (*see German* **abwechsel**); *lit.*, change off
wechselhaw, strike that changes targets
wechseln, *v.*, change
wechseln im durch, *idiom*, change one's enemy through, rotate around his weapon (*see Latin* **habitus mutatorius**)
weik, *see* **waik**
wenden, *v.*, turn, move
winden, *v.*, wind, curve, change; *w. weapons*; wind or rotate one's weapon around the opponent's weapon or limb, usually while maintaining pressure (*see Latin* **converso**)
wurff, throw
wurffen, *v.*, throw

Z

zeuchen, *v.*, withdraw, pull back
zu, *prep.*, to
zufechten, approach *(see Latin* **congressus***)*
zwifachen, *adv.*, double, twice
zwischen, *prep.*, between.
zwen, two
zucken, twitch
zornhaw, wrath strike
zwerchau, twitch strike

Appendix C

Literal Translations

SHORTSTAFF

Cod. Vinob. 10825, 154r
The first two upper binds from the right side, from the initial engagement.

With regard to this bind, position yourself by athletically wielding the staff in this manner: advance toward the enemy with the right foot, and in the initial engagement, bind him above from your right shoulder, with the right hand on the outside of his staff, and during this bind, feel whether he is holding the staff strongly or weakly. If the enemy strongly resists, then advancing with the left foot, and by winding the staff around his staff, from his left to his right side, be sure to thrust into the adversary's eyes.

However if he should use this technique against you during the first bind while you are facing the adversary with the right foot leading and your right hand forward, holding the middle of the staff, then parry his attack with your front point on the right side. Meanwhile step to the inside with the left foot, and thrust the rear point into his eyes. But if he sets that attack aside, then following with the right foot, smash the enemy's head from the bind with the forward point of your staff.

However, if he attacks your head from above, bring the left foot back, ward him off the front point, and be sure to immediately thrust into his chest. If the enemy parries, then thrust again, and aim for his right arm. At this point, you may safely withdraw from him in a defensive guard.

C93, 183r
The first two upper binds on the right side.

Note position yourself thus with your approach at the staff, step inside with your right foot towards him, and bind him above from your right shoulder with your right hand on the outside of his staff, in the bind feel if he is weak or hard in the bind, if he is hard against you, and holds strongly against you, then pass forward with your left leg, simultaneously change through from his left onto his right side, and thrust him to his face.

If he thrusts you to your face thus, and you are also standing against him in the bind [with] your right foot set in front, [with] your right hand placed forward in the middle of your staff, then set him aside with your forward point onto your right side, step inside with your left leg, and wind him [with] your back point to his face. If he sets you aside, then pass forward with your right leg, and strike him with your upper point out of the bind to his head.

If he strikes you above at your head thus, then set your left leg to the back, and then displace him with your forward point, simultaneously thrust him to his breast. If he displaces this, then withdraw your thrust again, and thrust him to his right arm, with this, move yourself back away from him into a good defense.

187

Cod. Vinob. 10825, 154v
The first two lower binds on the left side, from the initial engagement.

If you wish to use the aforementioned technique in this same way, then you will set the left foot in front; you will hold the spear above the left arm, so that your left hand is leading. Thus, if you take position against the adversary in this manner, and he in turn against you in the first lower bind of the staffs, then you will pass to the inside with the right foot, and from the inside thrust above his left arm on the outside; in fact, during this same thrust, surely drive your point into the adversary's chest. And if he parries this, advance again with the left foot, and curve the rear point of your staff toward the enemy's eyes.

If, however, the adversary uses the same thing against you, while standing with the left foot forward, then set his thrust aside, and following with the right, thrust the forward point into the enemy's eyes. However, if he parries, then by circling the staff around, thrust toward his left side. But if, on the other hand, he evades your attack, then advance with the left foot, and immediately attack the adversary's eyes with the front and rear point.

However, if he uses this same technique against you, then you will parry from the bind on both of your sides. Meanwhile, surely following with the right foot, thrust the point of your staff into his eyes. But if the enemy parries it, then with the right foot brought back, with the hand inverted, smash your staff into the enemy's head; quickly seize his staff with your left hand, and carefully guard your face with the most secure defense.

C93, 183v
The first two under binds on the left side.

Note, position yourself thus with this approach, stand with your left foot forward, and hold the staff on your left shoulder [with] your left hand in front, if you then stand opposing him thus, and he is also opposing you in the under bind, then step inside with your right leg, and thrust him from the inside out over his left arm, in this, thrust him [with] the point to his breast. If he displaces you, then again pass forward with your left leg, and wind him [with] your back point to his face.

If he thrusts you to your face thus, and you stand with your left foot forward, then set his thrust aside, and step inside with your right foot, simultaneously thrust him [with] your forward point to his face. Then if he sets you aside, then simultaneously change through, and thrust him to his left side. If he sets you aside, then pass forward with your left foot and wind him simultaneously quickly twice with your forward and back points to his face.

If he winds you twice thus, then set him aside with your bind from both sides, simultaneously pass forward with your right leg, and thrust in to his face. If he displaces you, then withdraw your right leg to the back, and strike him with your staff with an inverted hand to his head, simultaneously seize with your left hand again on your staff, and have care for your face with a good defense.

Cod. Vinob. 10825, 155r
The second two upper binds of the staffs from the left side

For the upper bind formed from the left side, you will thus position yourself for the purpose of properly wielding the staff. If the adversary takes position against you in the aforementioned manner––that is, the bind from the left side—then step to the inside with the right foot, and carefully determine whether he is holding the shortstaff stronger or weaker in the first bind. If you discover it to be weak, you will immediately follow with the right, and by thrusting, aim for his face. However, if he renders this attempt useless, then you will immediately adapt by rotating the staff from your right side into the adversary's right side.

On the other hand, if he uses the same thing against you, and you are standing in a position with your left foot forward during the first bind, then you will parry his attack; in fact, immediately advance with the right, and strike with your staff above his right arm into the enemy's eyes. When the adversary parries it, adapt by wheeling the point around into his right side, and thrust toward the right side. If the enemy parries it, then follow with the left, and by curving the rear point of your staff around, you will thrust between both of the adversary's arms toward the face.

However, if the adversary uses a thrust in this way, then you will bring your right foot back, and remember to parry his attack with the front point of your staff; meanwhile, you in turn will thrust the rear point toward his eyes, following again with the right, and smash your staff against his head with both hands, and in this manner you may withdraw from the opponent, in your strongest possible defensive guard.

C93, 184r
The second two upper binds on the left side.

Note position yourself thus with the approach into the upper bind on your left side, if he then stands also opposing you in the bind on his left side thus, then step inside with your left leg, simultaneously feel whether he is hard or weak in the bind, if he is weak, then pass forward with your right leg, and thrust him to his face. If he takes you aside, then simultaneously wind yourself through from your right [side] onto his right side.

If he winds you to your right side thus, and you also stand against him in the bind [with] your left foot set in front, then displace him, in this, pass forward with your right leg, and thrust in to his face over his right arm. If he displaces you, then wind yourself through with your point onto his right side, and thrust him with your point to his right side. If he sets you aside, then pass forward with your left leg, and wind him [with] your back point between both of his arms to his face.

If he winds you to your face thus, then set your right leg to the back, and set him aside with your forward point, simultaneously wind your back point also to his face, again pass forward with your right leg, and strike him with the half staff with both hands to his head, withdraw with this away from him [into a] good defense.

Cod. Vinob. 10825, 155v
Two lower binds from the right side.

When you want to correctly employ the lower bind, you will place the right foot forward, with the right hand directly next to the front point, and then rotate the staff upward, so that the right hand is positioned near the thigh, and aim for the adversary's face.

However, if he attempts the same thing against you, having also positioned yourself in a bind, and you place the right forward, then you should not hesitate to set this thrust aside with the forward point, and compose yourself thusly, as if you wish to attack his right leg with a thrust of the staff. Instead, following with the left foot, you will thrust the rear point toward either his eyes or chest.

On the other hand, when the enemy uses this same thing, you will step backward with the right, and parry the adversary's thrust with the rear part of the staff. But you also will not forget to pass to the inside again with the right foot, and thrust the front point into his eyes. However, if the enemy repels you, then you will wind the staff around his staff, and following with the left, strike toward his right flank with the aforementioned point.

And if the adversary attacks you in this manner, then parry it, immediately following with the right foot, and by rotating the staff, strike his face twice, once with each point. However, if he parries these, then you will remember to bring the right foot back, and thrust the staff toward his head. You will certainly take care when withdrawing from the enemy, so that you are not unguarded.

C93, 184v
The second two lower binds on the right side.

Note, position yourself thus with the approach into the lower bind, set your right foot forward, [with] your right hand extended forward near the point, simultaneously wind placing your right hand on your hip, and thrust him to his face.

If he thrusts you to your face thus, and you are also standing against him in the bind with your right foot set in front, then set him aside with your forward point, and as if you will want to thrust to his right leg, simultaneously pass forward with your left foot, and wind your back point to his face or his breast.

If he winds you to your face thus, then step with your right foot to the back, and set him aside with your back point, simultaneously step in again with your right foot, and thrust him with your forward point to his face. If he displaces you, then change through with your point, and pass forward with your left leg, thrust him with this on his right side.

If he thrusts toward you thus, then set him aside, in this, pass forward with your right leg, and wind him with both of your points twice to his face. If he sets you aside, then withdraw your right foot back, and strike him to his head, with this move yourself to the back into a good defense.

Cod. Vinob. 10825, 156r
A parry against the double middle-strike

For the purpose of the aforementioned technique, the staff must be wielded in an athletic manner; you will compose yourself in this way: you will step to the inside with the left foot, the front part of your staff should point toward the ground, the right hand should be positioned above the head, and at that moment, if you turn your staff by rotating upward, and you join the right hand to the thigh, then [you will] thrust the front point into the enemy's face.

However, if the adversary attempts the same thing, and you are standing against him in a position for the double middle-strike with the right foot forward, then parry his attempt with the front part of the staff; immediately following with the left foot, thrust the rear point of the staff into his face. But if he emerges safely from this technique, then by rotating below, you will batter his right arm. But if he frees himself again, then follow to the inside with the right, and thrust toward his eyes with the long point. However, when the enemy anticipates this thrust and parries, bring the right foot back, and attack the adversary's left foot with a strike.

But if that man attempts to attack with a thrust in the same manner, then you will not parry his attack; instead, by quickly thrusting, be sure to stab into his eyes, and then you will more carefully withdraw from the enemy.

C93, 185r
A displacement against a double middle-strike

Note, position yourself thus with the approach into this displacement, step inside with your left leg, your staff on the earth, your right hand extended out above your head, simultaneously wind yourself with your right hand onto your hip, and thrust him to his face.

If he thrusts you to your face thus, and you stand against him in the double middle-strike, [with] your right foot set in front, then set his thrust aside with your forward point, simultaneously pass forward with your left leg, and thrust your back point to his face. If he displaces you, then change through below, and strike him to his right arm. If he continues to displace you, then step inside with your right leg, and thrust him with your staff's point to his face. If the enemy notices the thrust and displaces it, then withdraw your right leg to the back, and thrust him to his left leg.

If he thrusts you to your left leg thus, then displace him thus, by specifically thrusting him quickly to his face, simultaneously move yourself with this from him to the back into a good defense.

Cod. Vinob. 10825, 156v
The left and right defense, or "fortress."

For the "fortress" from the right side, you will adjust yourself thusly: place the right foot forward; hold the staff at its strong point, from the right side. And if the adversary in turn stands against you at the strong point from his left side, the left leading, and you both bind the staffs after the first clash, then compose yourself, and you nevertheless want to attack the enemy's eyes with a thrust. But from this feint, rotate the staff by pivoting around from his right side to the left, and following with the left foot, attack the adversary's chest with a thrust. But if the enemy sets the thrust aside, then immediately follow with the right, and thrust the rear point of the staff between both of the adversary's arms into the eyes.

However, if he uses the same thing, then you will step to the inside with the right foot, and parry his thrust with the rear point; in that instant, be sure to change the staff around his staff by rotating with the long point, and you will again thrust it into his face. However, if the adversary parries it, then if you set the left in front, you will extend the forward point into the enemy's face; turn the body thusly, so that the right is leading, and smash the head with a strike of the staff.

On the other hand, if he in turn uses this same technique, then drive the staff to the ground from your right side, and if you advance against him, parry his strike between both of your hands; meanwhile, immediately attack his eyes with a thrust, and nevertheless withdraw from him in your strong defense.

C93, 185v
The right and the left guard.

Note, position yourself thus in this guard on the right side, stand with your right foot forward, and hold your staff on your right side at the strong. Then if he stands thus against you on his left side at the strong, with his left foot set in front, and having both one another tightly-bound, then you will be as if you want to thrust him to his face, simultaneously change through from his right onto his left side, passing forward with your left leg, and thrust him to the breast. If he displaces you, then pass forward with your right leg, and wind your back point inside in between both of his arms to his face.

If he winds you to your face thus, then step inside with your right foot, and set him aside with your back point, simultaneously change through with your point, and also thrust him to his face. If he sets you aside, then pass-over and strike him with a blow to his head.

If he strikes you to your head thus, then set your staff on the earth on your left side, and approach him with your staff, and displace him in between both of your hands, simultaneously thrust him to his face, and withdraw yourself with this to the back into a good defense.

Cod. Vinob. 10825, 157r
The thrust which attacks the face against a low parry with either point of the staff.

For this form, position yourself in this way: you will hold the staff on the right side; place the left hand in the middle of the spear, [with] the right hand near the right foot and the forward point of the spear extended toward the adversary's face. But if the opponent stands against you in a guard, the left foot leading and the front part of his spear driven to the ground, then with the right hand positioned directly adjacent to the rear point, following with the right foot, you will attack his eyes with a thrust.

On the other hand, if the enemy attacks you with the same technique, then you will step to the inside with the right foot, and during this attack, thrust your staff straight at his eyes, and parry the adversary's thrust; at that time, you will also wind the rear point of your staff between both of his arms and thrust toward the throat.

However, if he uses a similar technique against you, then step back with the right foot, and parry his thrust with the front point; you will also strike the enemy's head from the bind. Meanwhile, withdraw from the adversary in a strong defense; nevertheless, you will pay careful attention to his capabilities, so that he does not unexpectedly follow by pursuing you.

C93, 186r
A face thrust against a lower taking-off.

Note, position yourself thus with the approach into the face-thrust, hold your staff on your left side with your left hand at the half staff, your right hand back by your right leg, your forward point toward his face. If he stands then in the taking-off against you, [with] his left foot set in front, the staff toward the ground [and] the right hand extended back near the point, then pass forward with your right leg, and thrust him to his face.

If he thrusts you to your face thus, then step inside with your right leg, and going with the step, strike for his face with the staff, and set his thrust aside, simultaneously wind your back point in between both of his arms on-the-inside to his throat.

If he winds you to your throat thus, then set your right leg again to the back, and set him aside with your forward point, and strike him from the bind to his head, with this withdraw yourself to the back into a good defense, and take good care to watch your face lest he gets you by traveling after.

Cod. Vinob. 10825, 157v
Two parries from both sides.

When you want to properly employ the aforementioned technique, you will not hesitate to place your right foot forward; with the arm extended, tilt the staff toward the ground with its front point; the left hand should be placed in the middle of the staff, the right certainly high up by your face. And if the adversary stands against you in the same position on his left side, then you will step to the inside with the left foot, and wind the high point between both his arms toward the eyes. However, if he parries this attempt, then immediately following with the right foot, you will strike the right arm from the parry with the rear point of your staff.

On the other hand, if the enemy attempts the same thing, then having placed the left foot forward, you will set the enemy's attack aside with the forward point, away from your right hand; moreover, quickly following with the right foot, thrust the rear point of the staff into his chest. However, if he parries this attack, then follow to the inside with the left foot, and twice thrust both points of the staff toward his face.

On the other hand, if the enemy uses an attack in the same manner, then you will bring the right foot back, and be sure to carefully set the adversary's attack aside with the forward point; in that instant, you will thrust the rear point of the staff toward his eyes or chest, and withdraw from him in your secure defense.

C93, 186v
Two displacements from both sides.

Note, position yourself thus into this displacement with the approach, stand with your right foot forward, [with] your staff on the ground, with extended arms, your left hand in the middle of the staff, the right [hand] on top before your face. If he also stands thus identically against you on his left side, then step inside with your left leg, and wind your upper point in between his arms to his face. If he displaces you, then pass forward with your right right foot, and strike him from the displacement with your back point to his right arm.

If he strikes you to your right arm thus, and you are standing with your left foot set in front, then take it aside with your forward point in your right hand, simultaneously pass forward with your right leg, and thrust your back point to his breast. If he displaces you, then step inside with your left leg, and wind twice with your point to his face.

If he winds twice thus, then withdraw your right leg back, and set it aside with your forward point, simultaneously thrust him with your back point to his face or the breast, and withdraw yourself with this to the back into a good defense.

Cod. Vinob. 10825, 158r
A technique by which the eyes are attacked, against a parry.

For the purposes of this technique, you will compose yourself by wielding the spear into the engagement in this manner: you will set the left foot in front [and] hold the spear in the right hand from the right hip; the point should also be extended toward the adversary's face. However, if he happens to be strong, so that he positions himself against you in a parry, the right foot leading, and he holds the spear in the middle with both hands directly in front of the face, then you will pass to the inside with the right foot, and thrust the rear point of the staff into his face from the right side. However, if he parries, [then] immediately thrust the forward [point] into his chest; meanwhile, also bring the right foot back so that you thrust the staff into the adversary's right side.

But if he attempts to use this same method against you, [then] you will step toward him with the left, and be sure to parry his attack between your hands from your right side; also, quickly thrust the front part of the staff into the enemy's eyes. On the other hand, when he parries the same thing in a similar manner, [then] following with the right foot, aim for the adversary's groin with the rear point.

But if he attacks you below in this same manner, [then] you will step back with the left foot, and also parry his thrust with the rear point; in fact, you will immediately bring the left foot back to the inside, and attack his eyes with a double thrust; then, from this position, withdraw from the adversary while guarded.

C93, 187r
A face-thrust against a displacement.

Note, position yourself with the face thrust in the approach thus, stand with your left foot forward, your staff in your right hand by your right hip, the point toward his face. Then if he stands against you in the displacement with his right foot set in front, his staff in the guard, with both of his hands extended in front of his face, then step inside with your right foot, and wind your back point to the right side of his face. If he sets you aside, then wind your other point to his breast, simultaneously set your right leg to the back, and strike him with your staff to his right side.

If he strikes you to your right side thus, then step inside with your left leg, and set him aside in between both of your hands on your left side, simultaneously thrust him with your forward point to his face. If he displaces you, then pass forward with your right foot, and thrust in with your back point to his groin.

If he thrusts at you below in the same way, then step with your left foot behind his, and set it aside with your back point, simultaneously step inside with your left foot, and thrust him with a double thrust to his face, [then] step with this to the back into a good defense.

Cod. Vinob. 10825, 158v
Two attacks from which are formed a way of throwing the enemy down.

When you step within range of the adversary by athletically wielding the staff, you will advance to the inside with the left foot, and thrust the spear into the enemy's face from your right side. However, if the adversary parries it, [then] bring the right foot back, and if you withdraw the staff by rotating through the hands, smash his head with the long point.

But if the enemy uses the same technique, then you will parry his blow between your hands on the staff. Meanwhile, advance to the inside with the right foot, and thrust the leading point of the spear toward his eyes. When the adversary parries it, you will change by winding through his staff; following with the left foot, thrust into the enemy's right flank with the rear point.

But if he attempts the same thing, [then] you will parry the blow, and if you advance toward the adversary with the right foot, [then] aim for his face with the rear point.

However, if he attacks you with a similar technique, you will not hesitate to parry the enemy's strike, and step inside toward the enemy with the right, and during the parry, apply your staff's forward point to his throat, and if you kick the right foot into the enemy's knee, push up, [and] also bend down, you will throw the adversary.

C93, 187v
Two break-ins out of which goes a throw.

Note when you come with the approach toward the opponent, then step inside with your left leg, and thrust him with your staff to his face from your right side. If he displaces you, then set your right leg to the back, and loose your staff through your hands to shoot, and strike in with the long point to his head.

If he strikes you to your head thus, then displace him in between your hands on your staff, simultaneously step inside with your right leg, and thrust him with your forward point to his face. If he sets you aside, then change through on his staff, passing forward with your left foot, and thrust him with your back point to his right side.

If he thrusts you to your right side thus, then displace him, and step inside with your right leg toward him, simultaneously thrust him with your back point to his face.

If he thrusts you to your face thus, then displace him, [and] step inside with your right leg toward him, and in the displacement, fold him with your forward point on his neck, and with your right foot behind his left into the leg-hook, press up from you, and pull down toward you, then he will fall to his back.

Cod. Vinob. 10825, 159r
Two thrusts toward his upper openings.

If you advance toward the enemy while wielding the staff in an athletic manner, you will set the left foot in front, hold the staff straight in your arms, with your right hand positioned above your head, [and] the point also extended toward the adversary's chest And if he in turn stands against you in a high thrust, the left leading, [then] immediately follow with the right foot; by rotating the staff toward your right side with the right hand, after this rotation thrust the long point of the staff into his face. However, when he anticipates the thrust, and he evades it by defending himself, if you cross over by changing the staff from his left side to the right side, [then] attack the adversary's chest with a thrust.

And if he uses the same technique, [then] parry his attack, and follow with the right foot; meanwhile, forcefully thrust the rear point of your spear into the enemy's face, and if you also bring the right foot back during his own change-through, thrust the long point into his own chest. However, if he parries it, [then] step again to the inside with the right, and by striking with the forward point you will smash the adversary's head.

However, [if] he attempts that same thing, [then] you will set it aside by rotating the spear; meanwhile, be sure to attack by thrusting twice toward the face, and withdraw from the adversary in your strong defense.

C93, 188r
Two thrusts to his upper opening.

Note, when you come with the approach toward the opponent, then step inside with your left leg, and hold your arm extended on the staff [with] your right hand above your head, the point toward his breast. Then if he also stands against you thus in the upper thrust to your upper opening [with] his left foot set in front, then pass forward with your right leg, and wind with your right onto your right side, and in this wind, thrust your long point into his face. If he should notice the thrust and displaces you, then change through from his left onto his right side, and thrust him to his breast.

If he thrusts toward you thus, then set him aside, and pass forward with your right leg, simultaneously wind your rear point to his face, and in your wind, then step with your right foot to the back, and thrust your long point to his breast. If he displaces you, then again step back with your right foot, and strike him with the long point above to his head.

If he strikes toward you thus, then set him aside with a wind of your staff, simultaneously thrust twice to his face, and step with this to the back into a good defense.

Cod. Vinob. 10825, 159v
The long point against the first bind.

In the approach against the enemy after this strike, which is thrust toward him from the ear, you will remember to strike twice, and if you move closer toward him, [then] place the left foot in front, bring the right hand to the right hip, the point of the staff extended toward the enemy's face, [and] immediately following with the right foot, pierce his throat with the long point.

However, if he uses this same technique against you, [then] you will set the left in front, take position in the bind, block the enemy's attack with the forward point, and following with the right foot, attack his head with a strike. In fact, during this same strike, you will apply the staff to the enemy's throat by pivoting away from the forward point; afterward, set the left foot in front of his right foot, push forward above, and take great care that you render the adversary immobile with this technique.

On the other hand, if he attempts the same thing, [then] follow with the left foot, and parry his point with your staff, and at that moment, thrust the forward point toward the enemy's face. But when he skillfully sets your thrust aside, turn the staff around, and attack his face a second time.

However, if he attacks you with a double strike, [then] set him aside with the forward point, and thrust the rear point into his chest; consequently withdraw from him in your strong guard.

C93, 188v
A long point against a bind.

Note, when you go toward the opponent with the approach, then freely strike twice toward him inside from the plunge-strike, so you shall come for the opponent, then stand with your left foot forward, your right hand by your right hip, [and] your staff with your point toward his face, simultaneously pass forward with your right leg, and thrust him [with] your long point to his throat.

If he thrusts you to your throat thus, and you stand with your left foot set in front against him in the bind, then take it away with your forward point, simultaneously pass forward with your right foot, and strike in to his head, and in this strike, then wind your staff forward around his throat, step [with] your left foot in front of his right, and press over from you, and show whether you might have weakened him to win.

If he desires to weaken you thus, then pass forward with your left foot, and take away his point with your staff, simultaneously wind your forward point to his face. If he displaces you, then change through, and thrust him again at the aforementioned place, to the face.

If he thrusts you twice to the face thus, then set it aside with your forward point, and wind the back point to his breast, [then] withdraw yourself with this to back away into a good defense.

Cod. Vinob. 10825, 160r
An inverted strike against a parry.

For the aforementioned technique, you will position yourself in this way: strike toward the adversary by executing a double-strike extended from the chest, and if you move close to him, you should stand opposite the enemy with the right foot set in front. In fact, by quickly rotating the staff, with that hand inverted in such a way that, using the hand, the staff is spun around in reverse above [his] arm, so that the point is leading, you will smash his head, and if you grip the staff again with the left hand, [then] extend the forward point into his eyes.

However, if you stand against the enemy with the left in front, holding the staff with the hands extended, and the point turned toward the ground, then parry his strike and [his] thrust between both hands with the staff, and advancing to the inside with the right foot, strike the short point into his face with the right hand. If he evades it, [then] with the right set back, thrust the long point toward the enemy's face.

But when the adversary uses the same technique against you, [then] you will set both thrusts aside with the forward point, [and] immediately following with the left foot, thrust the rear point into the adversary's face twice. And if the enemy anticipates this attack and evades, [then] to such an extent as you are able, you will smash his right side with the long point [of the] staff. However, if he parries the strike a second time, then thrust the staff twice into his chest, and you may withdraw from the adversary while guarded.

C93, 189r
An inverted strike against a taking-off.

Note: position yourself thus with the approach to strike your "free-long," with a double breast strike in toward him, when you then come at him, then step inside toward him with your right foot, simultaneously loose the staff to quickly overrun, and strike him to his head with an inverted hand, then seize with your left hand again onto the staff, and wind your forward point to his face.

Then if you stand against him thus, with your left foot forward, with extended arms on your staff, the point toward the ground, then take his strike and thrust aside in between both of your hands on your staff, step inside with your right leg, and wind your short point in your right hand to his face. If he displaces you, then again step with your right foot to the back, and thrust your staff's point to his face.

If he thrust you to your face thus, then take it aside with your forward point, simultaneously step inside with your left foot, and thrust your back point twice to his face. If he notices the double thrust and displaces you, then quickly strike him with your long point to his right side. If he continues to displace you, then wind him twice to his breast, step with this to the back into a good defense.

Cod. Vinob. 10825, 160v
The strike by which the adversary's arm is pinned, against a strike by which the groin is attacked.

When you pass within range of the adversary by properly wielding the staff, you will advance with the left foot, and hold the staff with the right hand above the head; place the left hand lower on the middle of the staff, and in this manner thrust toward the enemy's left elbow so that from this position you will lock his arm.

But if he attempts the same thing against you, and if you position the staff with the right hand above your head, and you place the left foot in front, [then] if you parry by lifting the staff up high, attack his groin with the leading point, and with this thrust you will render his attempt useless. Then, you will advance to the inside with the right foot, and by winding the rear point of the staff, thrust toward the adversary's eyes.

However, if he attempts to employ a thrust in the same way, [then] you will parry him with the forward point, and if you hold the staff directly in front of your face, step toward the enemy with the right, strip the adversary's staff out of his leading hand with your rear point, and attack his face with the forward point. However, if the enemy parries it, [then] bring the right foot backward, and thrust the long point of the staff into his chest; then, if you accomplish this, you may safely withdraw from the adversary while guarded.

C93, 189v
A stress thrust against a groin thrust.

Note, when you come toward the opponent with the approach, then step inside with your left leg, and hold your staff with your right hand above your head, your left hand will be in front on your staff, and thrust him toward his left elbow, then you will take him to stress it.

If he thrusts at you thus to stress your left arm, and you also hold your staff above your head in your right hand, [with] your left foot set in front, then set it aside high upon your staff, simultaneously thrust toward his privates [groin], then drive in with your thrust, and then it is over, otherwise then follow inside with your right leg, and wind your back point to his face.

If he winds you to your face thus, then set it aside with your forward point, and hold your staff extended before your face, in this, step inside with your right foot toward him, and strike his staff out of his forward hand with your back point, and thrust him with your forward point to his face. If he sets you aside, then withdraw your right leg to the back, and thrust your long point to his breast, step with this to the back into a good defense.

Cod. Vinob. 10825, 161r
A defense or guard from the balance, against a violent strike.

There are several guards from the balance. The first is, if you hold the rear point in front of the face with the right hand, the front [point] driven toward the ground, and the left hand in the middle of the staff, [then] you will turn the rear [point] on the right side. The second is in response to the first. Therefore, you will position yourself in that sort of guard, in such a way that you stand equally on both legs, the staff touching the ground; you will place the left hand in the middle of the staff, with the right certainly positioned at your right side, and compose yourself in a balanced stance. At that same time, you should definitely advance to the inside with the left foot and thrust the long point toward the adversary's chest.

However, if the enemy uses the same thing against you, and you place the left foot in front, the staff thus brought into contact with the left foot, so that its point touches the ground, the right hand raised above the head near the other point of the staff, then / step with the right and parry his own attempt with the forward point using the right hand; furthermore, quickly pass backward with the right foot, / and with the long point extended, use all [your] strength for a violent thrust toward his face.

Moreover, if the enemy attempts the same thing against you, you will pass with the left foot in a triangle, dodge his thrust, quickly proceed inward with the right, and by rotating the staff you will smash his head, and at that moment, with the left hand again gripping the staff, move away from the adversary by side-stepping twice.

C93, 190r
A guard in the scales against a forceful thrust.

Note, positioned thus with the approach into this guard, stand with equal feet together, your staff at the earth, your left hand in the middle of the staff, the right [hand] on your right side, and place yourself with your body into the "scales," simultaneously step inside with your left leg, and thrust your long point to his breast.

If he thrusts you to your breast thus, and you stand with your left foot forward, your staff on your left leg at the earth, your right hand above your head by your point, then step inside with your right leg, and set it aside with your forward point [with] your right hand, in this, step with your right foot again to the back, and thrust him with the "forceful" thrust, with your strength to his face with your long point.

If he thrusts you with strength to your face thus, then step with your left in a triangle, then you will go in from his thrust, simultaneously step inside with your right leg toward him, and loosing [releasing] your staff, over-run and strike to his head, in this, grasp with your left hand again onto your staff, and move yourself with this twice from him to the back into a good defense.

Cod. Vinob. 10825, 161v
A parry with the long point, against a crossed change.

When you approach the enemy by wielding the staff athletically, you will bring the right foot to the inside, with the front [of the] staff tilted toward the ground, bring the right hand to its middle, and place the left near the rear point, above the left hip.

However, if the adversary stands against you in a crossed-change guard, the right foot in front, [then] with the long point raised, and following with the left, aim for his eyes with a thrust. If the enemy evades it, [then] following with the right, you will thrust the rear point toward the adversary's face or chest.

However, when he uses the double-thrust from your previous attack against you, [then] with the staff raised, you will set his thrust aside with the forward point. In addition, following with the left foot, you will thrust the rear point toward his face from the crossed-change. However, if the enemy parries your thrust by fiercely defending himself, [then] to such an extent as you are able, you will attack his left flank twice by winding.

But if the adversary threatens you in this manner, [then] you will set his strike aside with the middle of the staff; immediately following with the left, thrust the rear point toward the adversary's left flank; from this position, bring the left foot back, and violently smash his head with the forward point; when this has thus been accomplished, you may withdraw from the enemy.

C93, 190v
A long point with a taking-off against the "crossed changer."

Note: when you come toward the opponent with the approach, then step inside with your right foot, and hold your staff forward at the earth, your right hand in the middle of it, the left back near your point by your left hip. Then if he stands against you in the "crossed changer" [with] his right foot set in front, then go up with your long point, simultaneously pass forward with your left leg, and thrust him to his face. If he sets you aside, then pass forward with your right foot, and wind your back point to his face or the breast.

If he thrusts at you twice to your face thus, then go up with your staff, and set it aside with your forward point, pass forward with your left leg, and thrust your back point to his face from the "crossed changer." If he displaces you, then wind quickly against him twice to his left side.

If he travels after you thus, then set it aside with your half staff, pass forward with your left leg, and thrust him with your back point to his left side, simultaneously withdraw your left leg to the back, and strike him with your forward point to his head, stepping with this, withdraw into a good defense.

Cod. Vinob. 10825, 162r
Two thrusts to the chest from above, both formed from the left side.

For this technique, you will position yourself in this way: advance toward the enemy with the left foot, and thrust the staff from your right side toward the chest, between both of his arms on his left side.

But if he in turn stands against you in the position of a high thrust from his left side, the left foot leading, and he attacks your chest in this same way, then you will remove the left hand from your staff and take hold of the adversary's staff near the point; curve the point of your staff beneath his left armpit with the right hand, advance toward the adversary with the right foot, and if you turn to the right side with the arms crossed and both staffs grasped, you will trap him in such a way that he can do absolutely nothing to you.

However, if he traps you with the same technique, [then] you will remember to let go of the staff, place the right on the outside behind the adversary's left foot, seize his right knee with the right hand, [and] also bring the left hand under his right armpit while maneuvering around the enemy's body, and in this way you will throw him away from your body to [his] defeat.

C93, 191r
Two upper thrusts to the breast from either left side.

Note, position yourself thus with the approach, step in toward him with your left foot, and thrust him with your staff from your left side to his left breast in-between the inside of his arms.

Then if he stands against you also in the upper thrust from his left side, and has his left foot set in front, and thrusts you also to your breast, then remove your left hand from your staff, and seize with it onto his staff by his point, simultaneously wind him with your right hand your point under his left shoulder, and step in toward him with your right foot, wind yourself with this onto your right side with crossed arms with both staffs, then blocked in so that he cannot come to work.

If he guards you thus so that you cannot come to work with your staff, then let your staff quickly fall, set your left foot in quickly behind his right, and grab with your right hand to his right knee-bend [back of the knee], and with your left under his right armpit, securely over around the body, then you will throw him so that he cannot cause you any harm.

Cod. Vinob. 10825, 162v
A throw by which he is restrained

You will compose yourself in this way: from the blow which is struck toward the enemy from either ear, you should strike, and if you move in closer [and] place the left foot in front, [then] attack the adversary's face with the forward point.

However, if the enemy uses the same thing against you, and if you approach by wielding the staff against [him] athletically, then you will set his strike aside by changing from the forward part of the staff, and advance toward the enemy with the right foot, and also attack his chest with the long point.

However, if he uses the same method against you, [then] move close to the adversary, and parry his thrust between your hands with the middle of the staff. Meanwhile, you will be sure to position yourself in a balanced stance while releasing the staff backward over the head, grab the adversary's groin with both hands, and flip the enemy upside down, and simultaneously lift him up high, [so that] he is made unable to fight back. Definitely knock the enemy over with this technique, and if he is prostrate, then press the right knee between his feet and into his groin, reach under the adversary's right with the left, and seize either his hands or throat, and if you press down strongly on every part of him, you will be able to hold him down.

However, if this is done to you by the adversary, [then] you will take care from the start that your hands are not captured. Moreover, you will quickly seize the face with one hand, so that the thumb is placed under that man's chin; press the other fingers of that hand to his eyes, [and] grip [him] firmly. Then, thrust the other hand into his groin with a powerful strike, [and] thrust out the foot which is freer, but then quickly bring it back; you will damage his groin the most from this technique, and with these three techniques you will defeat the adversary.

C93, 191v
A throw to a capture that he cannot arise from.

Note, hold yourself thus with this approach, strike yourself into the plunge-strike to the inside, set your left foot forward, and thrust him with your forward point to his face.

If he thrusts you to your face thus, and you are also opposing him in the approach, then set it aside with a forward wind on your staff, step inside toward him with your right foot, and thrust him with your long point to his breast.

If he thrusts you to your breast thus, then you will want to step inside toward him, and set his thrust aside in the middle of your staff in-between both of your hands, simultaneously throw your staff out over your head, and give yourself with your body into the scales, and seize with both hands to his weak, press in there, and lift up well over-himself, then you have weakened him, simultaneously throw him under you. If you have then thrown him, and to trap him in the stocks, then knee him with your right leg in-between both [of his] feet into his privates, and with the left under his right, and seize both of his hands or the throat, pressing almost everything to the earth.

[If] then to hold on to you, then he forces and throws you down, then having arrived, [it will be] good [if] from there your hands will not be locked; to separate from him immediately, use a hand on his face, the thumb on under the chin, the other finger below the eyeball to grip [it] strongly, with the other hand thrust in strongly striking to the privates, use the other foot that you have free, and withdraw it in quickly again to yourself, then give him a good one to the groin; with these three grips at this time may you escape from there.

Cod. Vinob. 10825, 163r
A parry with the middle of the staff from either side, with a thrust.

By athletically wielding the staff into the fight, compose yourself in this way. If the adversary stands with the right foot leading, using a thrust against you, and he attacks your eyes with this thrust [as depicted by the left figure], then prepare yourself thusly: set your right foot in front, and with both arms raised up high [as depicted by the right figure], observe the enemy, quickly move toward him with your left foot, and thrust your rear point into his eyes. You will parry his thrust in this way, and you will be able to use this technique on both sides, with either the middle of the [weapon] or the entire staff, by rotating the staff in a circle, [and] threatening with either the front or rear [point].

Cod. Vinob. 10825, 163r
A half staff to both sides from the approach to thrust with.

Note, hold yourself thus with the approach, if he stands with a thrust opposing you with his right foot forward, and thrusts you to your face, then position yourself thus, stand with your right foot set in front and with both arms up high and step with your left foot quickly toward him, and wind him with your back point to his face, then you will take his spear aside, and because you can use this to strike to both sides with the whole or the half staff, [change] through thus winding the staff to him backward and forward [illegible].

Cod. Vinob. 10825, 163v
A strike with a parry

    Prepare yourself thusly for this technique: strike your high strike with the long point, by aiming for his head, and follow with your left foot.

    If the enemy attacks you by striking in this way, [then] hold the staff exceptionally high in the air for a middle strike and a thrust, and advance toward the enemy with your left foot; block his strike or thrust with the middle of the staff by parrying in this way.

    And thrust your rear point between both of his arms into his eyes, and with the forward point, strike the staff into his head; by taking care while withdrawing from the enemy, in your defensive guard, you will not be found open.

Cod. Vinob. 10825, 163v
A strike with a displacement

    Note, hold yourself thus in this technique's approach: strike with the long staff from above to his head, and pass forward with your left foot.

    If he strikes toward you thus, bring your staff to a half thrust, well in the high, and step in on toward [him] with your left foot, and with this set his strike aside on the middle of the staff, and wind him with the back point inbetween both of his arms to the face, and with the forward point strike him to his head, and withdraw away from him with a good defense.

Cod. Vinob. 10825, 164r
An upper and lower bind using half-staffs.

If you bind him from from above, and he binds you from below, then advance with your left foot and thrust your back point into his own eyes. If he sets you aside, then strike his head with your front point.

If he strikes your head thusly, then set him aside with the half or whole staff, and quickly thrust or strike to whatever part of his body you can.

Cod. Vinob. 10825, 164r
A upper and lower bind to the half staff.

Note, hold yourself thus in this approach, bind over him thus from below and above one [to the] other, then step inside with your left foot, and thrust him with the back point to his face. If he sets you aside, then strike him with the forward point to the head.

If he strikes at you thus, set him aside with the half and the whole staff, and thrust and strike quickly at him wherever you may attack him.

Cod. Vinob. 10825, 164v
A parry with a strike.

You should place your left foot forward, and follow with the right, and employ a thrust toward the enemy's eyes, then parry his blow by setting aside in this way.

But if that man uses the same technique against you by setting aside, then advance with your left foot and thrust the shorter point into his eyes. If he parries this, [then] quickly move behind him, and hit his head, and quickly seize your staff again, and thrust a strike into his eyes with the long staff.

Cod. Vinob. 10825, 164v
A displacement against a blow.

Note, position yourself thus in this approach, stand with your left foot forward, and pass forward with your right, and thrust him to the face, thereby you will take his staff away.

When he takes your strike away thus, then pass forward with the left foot, and wind him from below in to his face. If he displaces you, then spring crosswise onto the left, and strike him to the head, and seize quickly to your staff again, and thrust with the long staff to his face.

LONGSTAFF

Cod. Vinob. 10825, 166r
The first two binds of the lances from the left side, using their stronger and weaker parts.

By wielding the lance athletically, compose yourself in this manner: you will hold the weapon from the left side, place the left foot forward, extend the lance's rear point from the right side with the right hand, and place the left hand on your lance on the left side. If you thus position yourself, you will advance to the inside, and set the right foot in front, and attack the adversary's face with a thrust.

However, if he attempts to use the same thing against you, with the left foot leading, and [then] and you grip your lance with the left hand near your left knee, and also with the right hand near the right foot, at the other point; set the enemy's thrust aside with the front part of the lance. Meanwhile, you will certainly rotate the lance by winding, and following with the right, attack the enemy's chest from [your] right side into his right flank. However, if he sets this thrust aside by defending himself, [then] you tilt the front part of the lance toward the ground, step back with the right foot, with your lance positioned in front of the face, and lift the right hand above your head away from the ground, and furthermore, quickly bring the right [hand] holding the lance toward the right side; if you advance against the enemy again with the right, [then] thrust the lance from the right side toward the adversary's left, into his face.

However, if the enemy attacks you in this way, [then] parry it by winding, and bring the right foot back; if you raise the lance up high, [then] transition to the strong part; that is, extend the lance into the enemy's eyes again, and pay attention to whether he resists strongly or weakly, on the right side, and once again grip the lance in the front part, and thus following with the right foot, thrust [your weapon] from the adversary's right side into his face. On the other hand, if the enemy parries it, [then] step backward with the right, and meanwhile tilt the front part of the lance toward the ground, and with [the lance] raised high above your head in your strong guard, withdraw from the adversary.

C93, 194r
The first two binds from the left side with the weak and strong.

Note: position yourself with the approach from your left side with your long spear thus, set your left foot forward, and hold your back point in your irght hand, be sure to extend [it] out on your right side, [with] your left hand on your left side upon your spear, simultaneously step inside with your right leg, and thrust him to his face.

If he thrusts you to your face thus, and you are standing with your left foot forward, your left hand by your left knee upon your spear, your right hand at your right leg, your point at [his] face, set his thrust aside on the front of your spear, simultaneously change through, and step back with your right foot, and thrust him to his breast from your right onto his right side. If he sets you aside, then set your spear forward on the earth with your point down, simultaneously step with your right leg to the back, your spear in front of your face, with your right hand above your head, simultaneously wind your right hand onto your right side with your spear, also step inside with your right foot, and thrust him from your right onto his left side, to his face.

If he thrusts toward you thus, then wind him aside from your spear, simultaneously step with your right leg to the back, and swing your spear at him, then step into the strong on your right side, and seize with your left hand again forward on your spear, pass forward with your right foot, and thrust him on your right side into his face. If he displaces you, then set your right foot again to the back, simultaneously loose your spear forward by the point down onto the earth, and wind yourself on into a good defense with your spear above your head, and withdraw yourself with this to the back.

Cod. Vinob. 10825, 166v
The strong and weak parts of the lance from the first bind.

Next you will approach the enemy by properly wielding the lance; you should pass to the inside toward him with the right foot, and with the lance directed toward his face, the left hand should be positioned above the left foot, and from this position, if you advance to the inside toward the enemy with the left foot, [then] attack his throat with a thrust of the lance.

However, if the adversary has attacked you from the same position, the right foot leading, with your right hand placed near the right knee, and the with left hand positioned above the left foot, [then] you will grip the lance, and having now positioned yourself according to [this] description, parry his thrust from your right side using the front part of the lance; and if you quickly advance to the inside with the right foot, attack his eyes. But if the enemy parries your attack, [then] aim for the chest with a thrust by changing the staff underneath. Meanwhile, certainly following with the right foot, by raising the lance in a balanced stance, he is checked; if you hold the lance near either the left or right side, bring [it] toward the head, and aim for the adversary's eyes with a thrust of the lance.

And if the enemy attacks you from this same position, [then] bring the left back, and evade his [attack] with a crossed-change toward your left side, and take care that you also bring the right back, and from the change you will wind the lance up high toward the right side, set the right foot to the inside again, and violently thrust at his eyes. But if the enemy should render your attempt useless, [then] if you bring the lance back, you will aim for the aforementioned part of the enemy's face. Meanwhile, you will also let the front part of the lance drop downward, certainly lift the right arm high above the head, and hold the lance in front of your face [to form] your strong guard.

Moreover, if the adversary uses a double thrust of the lance against you, then you should not hesitate to parry his thrust between both of your hands; following to the inside with the left foot, and having attacked his chest with a thrust of the lance, take care that you withdraw from the adversary while guarded.

C93, 194v
The weak and strong from the bind.

Note, when you come toward the opponent with the approach, then step inside with your right foot toward him, hold your spear toward his face, with your left hand over your left leg, simultaneously step inside with your left foot, and thrust him to his throat.

If he thrusts toward you thus, and you stand with your right foot forward, [with] your right hand over your right knee, [with] your left hand over your left leg on your spear, then set his thrust aside [with] the front of your spear onto your right side, simultaneously step inside with your left leg, and thrust him to his face. If he displaces you, then change through under it, and thrust him to his breast, simultaneously pass forward with your right foot, and go up with your spear into the "free stance", and thrust him to his face.

If he thrust toward you from the "free stance" thus, then withdraw your left leg to the back, and set him aside with [the] "crossed changer" onto your left side, simultaneously step with the right foot also to the back, and wind yourself up from the changer onto your right side with your spear, in this, again step inside with your right foot, and thrust him with the strong to his face. If he displaces you, then withdraw your spear again to you, and thrust him quickly again to the aforementioned place, his face, simultaneously let the front of your spear down with your right arm well up over your head, your spear before your face in a good defense.

If he thrusts you twice to your face, then take his thrust away in-between both of your hands on your spear, simultaneously pass forward with your left leg, and thrust him to his breast, move yourself with this to the back into a good defense.

Cod. Vinob. 10825, 167r
The right and left open guard

For the purposes of the right open guard, you will position yourself in this way: you will place the right foot forward, the forward part of the lance having been tilted toward the ground, the left hand placed near the rear point, and the right near the right knee, and at the same time advance to the inside with the left foot, raise the forward point up high, and aim for the adversary's eyes.

However, if he attempts the same thing against you, [with] the left foot leading, and your left hand gripping the lance on the left knee, with the right hand definitely placed near the rear point, then you will set the adversary's thrust aside from the open guard in a crossed guard, and bring the left foot backward. Meanwhile, however, you will wind upward into a free stance on both legs, so that the lance is resting on your left elbow; immediately following with the left foot, violently attack the enemy's eyes with a thrust. And if the enemy parries this attempt, [then] follow with the right, [and] aim for the chest from his left side.

But if he attacks you with the same technique, then [parry] his own thrust with the front point of your lance, from your left side toward your right; moreover, immediately following with the right foot, you will attack the adversary's groin with a thrust. When the enemy renders your attempt useless, follow with the left, with the lance raised high above the head, [and] crush his face with a thrust.

However, when he employs the double-thrust, you will parry him with the forward point, immediately thrust into his eyes or chest by winding the lance from the adversary's left side toward his right, and then following with the right, you will be able to strongly thrust the lance toward the adversary from a distance, and from this position, position yourself in an open stance on your right side, so that the right foot may be turned backward during this same transition, and in this way withdraw from the enemy while in your strong guard.

Dresden 195r
The right and the left open guard

Note, position yourself thus with the approach into the right open guard, stand with your right foot forward, with the front of your spear toward the earth, [with] your left hand by your rear point, [with] your right [hand] by your right knee, simultaneously step inside with your left leg, go up with the front of your spear, and thrust him to his face.

If he then thrusts you to the face thus, and you are standing with your left foot set in front, [with] your left hand forward by your left knee, [with] your right hand on your spear by your point, then set his thrust aside from the open guard into the barrier guard, and set your left foot to the back, simultaneously [with] yourself again in a free stance with equal feet together, so that you carry your spear in upon your left elbow, then pass forward with your left leg, and thrust him to his face. If he displaces you, then pass forward with your right foot, and thrust him to his left breast.

If he thrusts toward you thus, then take his thrust aside with your long point from your left onto your right side, simultaneously pass forward with your right leg, and thrust him to his privates. If he displaces you, then pass forward with your left foot, going up with your spear well over your head, and thrust him to his face.

If he thrusts twice toward you thus, then set him aside with your spear's forward point, simultaneously change it through from your left side onto his right to his face or the breast, and pass forward with your right leg, then you [will] quickly do a long thrust, in this, wind yourself into the open guard on your right side, so that your right foot goes to the back with the wind, and move yourself with this again to the back into a good defense.

Cod. Vinob. 10825, 167v
The free stance against the crossed guard

When you advance toward the enemy while flourishing the staff in an athletic manner, you will position yourself in a free stance in this way: you must stand in an upright position of the body in such a way that the left side is leading, the lance is extended toward the adversary on the left arm, and the right hand is placed at the rear point; following at once with the right foot, thrust the lance toward the enemy's body with all your might.

However, if he attacks you in this same way, standing in the crossed guard and [your] right foot facing the adversary, you will remember to bring the right backward, and parry his thrust; with the enemy's lance thus set aside, you will turn to the right side. In fact, during this same parry, if you follow with the right foot, stab into the adversary's eyes with a thrust. But if he parries your thrust, [then] follow with the left, and thrust the staff into the his chest by winding from your right side toward the enemy's left.

However, if the adversary attempts this same thing against you, [then] you should not at hesitate to parry his attack with the forward point of the lance, and meanwhile, following with the left foot, you will crush the enemy's chest by thrusting. But if he blocks this thrust, then with the lance wound from the right side toward his left, you will aim for that same target [again].

However, if he attacks you twice using this same technique, the left foot placed in front, you will parry that thrust with the forward part of the lance, and immediately follow with the right foot, and if you have attacked his eyes, [then] you will bring the right foot back, tilt the forward point of the lance toward the ground, and with the right hand raised above your head, you should take care to hold the lance in front of your face. If it turns out that the adversary follows by pressing himself forward, then thrust the lance toward him as far as you can with the long [point], and step back into your strong guard.

C93, 195v
The free stance against the barrier guard

Note when you come toward the opponent with the approach, then hold yourself thus with the free stance, stand erect with your body, [with] your left side turned-forward, [with] your spear toward the opponent on your left arm, [with] your right hand extended back by your point, simultaneously pass forward with your right leg, and thrust him violently to his body.

If he thrusts toward you thus, and you stand in the crossed changer, [with] your right foot set in front against the opponent, then step back with your right leg, and set his thrust aside, turn yourself with this onto your right side, and into the displacement, then pass forward with your right foot, and thrust him to his face. If he displaces you, then step back with your left leg, and change through from your right onto his left side, with this thrust him to his breast.

If he thrusts up at you thus, then set him aside with your forward point [of] your spear, simultaneously pass forward with your left leg, and thrust him to his breast. If he sets your thrust aside, then go through below from his right onto his left side, and with this thrust him to the same place.

If he thrusts twice toward you thus, and you stand with your left foot forward, then set his thrust aside forward on your spear, simultaneously pass forward with your right leg, and thrust him to his face, and withdraw your right leg again to the back, [with] your spear forward on the earth, and then with your right hand well on over your head, so that the spear is placed before your face. If he travels after you, then thrust a long thrust up in with your spear, [then] move yourself with this to the back into a good defense.

Cod. Vinob. 10825, 168r
A bind of the spears from the middle against a thrust that attacks the enemy's eyes.

For this technique, you will compose yourself in this way: you will set the left foot forward, place both hands in the middle of the lance, the left should still be leading, with the right positioned near the left foot, and from this position, thrust the lance through the hands into the enemy's chest.

But if he uses this same technique against you, during a thrust that is aimed at the eyes, standing upright, and the left foot leading, [then] follow with the right; turn the lance toward your left side; and from this position, stab into the adversary's eyes. However, if he attempts to parry, then bring the right foot back. But if the adversary threatens you by advancing [and] thrusting the lance toward you, then you will turn yourself into a crossed changer on your right side, and parry his own thrust. On the other hand, if you advance toward him with the right on the inside, [then] attack the enemy's chest or eyes with a thrust of the lance.

But when he in turn attacks you with a double thrust, you will parry him with the forward part of the lance, and at the same time, by winding the lance from your right side toward the adversary's left, you will smash the staff into his own left with a thrust. When the adversary parries [your thrust], following with the left foot, stab into his groin.

When he actually uses this same technique against you, [your] right foot leading, you will remember to bring [the right foot] back, and evade the enemy's thrust by parrying on your left side; in fact, immediately step forward again, and aim for the enemy's leading foot with the point of the lance. However, when you want to escape from this attack, with the staff quickly lifted up, and pointed toward his own eyes, you will withdraw from him backward in your strong guard.

C93, 196r
A bind with the half spear against a face thrust.

Note, position yourself thus with your approach into the bind with your half spear, stand with your left foot forward, with both hands in the middle of your spear, so that your left hand is forward, quickly set the right by your left leg, simultaneously shoot your spear through your hand[s] to his breast.

If he thrusts toward you thus, and you oppose him in the face thrust, stand with your left foot forward, then pass forward with your right leg, and wind with your spear onto your left side, and thrust him with this to his face. If he displaces you, then withdraw your right leg again to the back. If he travels after you with a thrust, then move yourself into the crossed changer on your right side, and take his thrust away, simultaneously again step inside with your right leg, and thrust him to his face or the breast.

If he thrusts twice toward your face thus, then set him aside forward on your spear, simultaneously change it through from your right side onto his left, and thrust him to his left shoulder. If he displaces you, then pass forward with your left leg, and thrust him to his privates.

If he thrusts below toward your privates thus, and you stand with your right foot in front, then set it to the back, and turn his thrust aside onto your left side, simultaneously quickly step inside again, and thrust him to his leg that is set in front. If he wants to displace your thrust, then go quickly again up in to the face, and withdraw yourself with this to the back into a good defense.

Cod. Vinob. 10825, 168v
A thrust by which the eyes are attacked, against [a thrust] by which the groin is attacked.

When you arrive at the enemy by wielding the lance athletically, you will pass to the inside with the left foot; and from your left side toward his right, crush the adversary's eyes with the point of the lance.

However, if he assails you in this same way, standing with the left foot toward him, [then] drive him off with the forward point. Certainly attack his groin at the same time.

If the enemy uses the same thing against you, [then] bring the left foot back, and parry his attack with the forward part of the lance; moreover, if you set the left foot to the inside again, [then] attack the chest with the point of the lance. However, if he parries the thrust while defending himself, [then] you will wind your [weapon] up away from his lance straight toward the enemy's eyes.

But if he employs this same double strike, then you will not hesitate to set him aside with the middle of the lance. Also, following simultaneously with the right foot, you will again batter his eyes with the point of the lance. But if he notices this thrust and parries, [then] having rotated the lance, follow with the left foot, and crush the adversary's right flank with the point of the lance.

However, if he attempts this same thing against you, the left foot leading, then from your right side you will parry the enemy's attack with the forward part of the lance; during this same parry, certainly step to the inside with the right foot, and stab from the adversary's right side into the chest. When the enemy parries, you should bring the right foot back, and position yourself with the lance in a powerful stance; that is, you should extend the lance toward the adversary's face, and feel if he is holding [his weapon] strongly. Thus, from this position you will struggle with the adversary, by distinguishing one opening from another, first high, then inward, wherever the opening may be.

But if the enemy seeks out your openings with this same technique, [then] with the lance raised above your head, you will parry his attack; meanwhile, also attack his eyes, and withdraw from him; make sure that this is done while protected.

C93, 196v
A face thrust against a groin thrust.

Note, when you come toward the opponent with the approach, then step inside with your left leg, and thrust him from your left side onto his right to his face.

If he thrusts you to your face thus, and you stand opposing him with your left foot forward, then set his thrust aside with your forward point, simultaneously thrust him toward his privates.

If he thrusts to your privates thus, then set your left leg to the back, and set his thrust aside forward on your spear, simultaneously again step inside with your left leg, and thrust him to his body. If he displaces you, then wind upon his spear straight up to his face.

If he then thrusts twice toward you, then displace him [with] the middle of your spear, simultaneously pass forward with your right leg, and thrust him also to his face. If he notices the thrust, and if he then sets you aside, then change through, passing forward with your left foot, and thrust him to his right side.

If he thrusts you to your right side thus, and you stand with your left foot forward, then set his thrust aside forward on your spear onto your right side, and in this displacement, step inside with your right leg, and thrust him to his breast on his right side. If he displaces your thrust, then step again with your right foot to the back, and give yourself with your spear onto the strong, simultaneously work with him from one opening to the other, below and above, wherever you may find it.

If he seeks your openings thus, then go with your spear well up over your head, and set him aside, simultaneously thrust him to his face, and withdraw with this to the back into a good defense.

Cod. Vinob. 10825, 169r
The "simple changer" against a guard from the half-lance.

You should compose yourself in the "simple changer" as follows: you will set the left foot forward, the point of the lance tilted toward the ground, the left hand holding the lance near the left knee, [and] the right placed by your right side, near the rear point, and from this position, following with the right foot, you will pierce the adversary's left flank.

But if the enemy tries the same thing against you, standing in a half-lance guard, and the left foot leading, and with both hands holding the lance near the chest on the right side, [then] following with the right foot, you will parry by setting the enemy's attempt aside with the middle of the lance. Meanwhile, you should also advance toward him to the inside, and thrust the half-lance toward his eyes; thus, if you employ this, all his efforts will be in vain.

However, if he pursues you by closing in and pressing forward in this same way, and you are able to accomplish nothing with the forward point because of the enemy's approach, then if you wind the lance backward by rotating through the left hand, you will be able to work again and fight with the adversary, and you in turn will rush toward him, and strike the enemy's closest opening with a half-lance technique, wherever an opening appears. But if he parries it, [then] following with the left foot, you will jab the adversary's chest.

However, if he seeks out your openings with this same technique, then step back twice into a "crossed changer" guard; bring the right hand as far back on the lance as you possibly can, near the rear point, and fight with the enemy from the strong and weak position by seeking out one opening after another. If he evades your thrusts again by parrying, [then] wind the lance far from the right hand toward his eyes; meanwhile, also retreat backward as safely as you can.

C93, 167r
The simple change against a guard in the half spear.

Note, [to] position yourself with your approach into the simple change, stand with your left foot forward, with your spear point on the earth, [with] your left hand by your left knee on your spear, [and] your right by your right side back by your point, simultaneously pass forward with your right leg, and thrust him to his left side.

If he thrusts to your left side thus, and you stand in the guard with your half spear, [with] your left foot set in front, your spear with both hands in front of your breast on your right side, then step back with your right leg, and set him aside with your half spear, simultaneously spring inside toward him, and let your half spear shoot to his face, then he will not want to come to work.

If he travels after you thus, and you can work with your forward point, then let your spear shoot to the back through your left hand, then you may with this come again to the same work, in this, also spring inside again toward him, and work with your half spear to his next opening, where ever it may be. If he sets you aside, then pass forward with your left foot, and thrust him to his body.

If he seeks your openings with his half spear thus, then withdraw yourself twice to the back into the "crossed changer", and take your spear again in your right hand back by your point, upon the full-length, simultaneously work with his weak and strong from one opening to the other, where ever it may be. If he displaces you, then thrust him with a long thrust with your right hand to his face with your spear, and withdraw yourself with this to the back into a good defense.

Cod. Vinob. 10825, 169v
An inverted thrust against a parry.

For this method, which will be revealed in the following, you will position yourself for the inverse strike in an athletic way. You will place the left foot forward, the left hand holding the lance near the left knee, the right definitely placed between both feet near the rear point, so that the forward point is extended toward the adversary, and with you standing thusly, follow with the right, and position yourself as if you wish to strike his left side; instead, with the hands turned upside-down, or "changed," strike the lance into the enemy's right side from the inverted thrust.

However, if he attempts the same thing against you, standing in a parrying position, you will set the left foot in front, tilt the front part of the lance toward the ground, and raise the rear point in front of the face with the arms straight, then from above and below, from one side to another, you will parry his attack. When he is thus pushed back, follow with the right foot, and with the lance wound up high on your right side, crush the adversary's throat with a thrust. But if he defends himself from this thrust, having rotated the lance from your right side toward the enemy's left, aim for his own eyes. However, if he anticipates it and parries, [then] following with the left foot, attack his left side with a thrust of the lance.

If, however, he employs this same technique against you, leaning on the right, then if you change the right hand to the left side, the enemy's thrust will be blocked. Meanwhile, you should also bring the right foot back, and position the lance on your left side again, at the strong point, and following again with the right, if you attack [his] upper opening with a thrust, [then] retreat from the adversary in your protected guard.

C93, 197v
An inverted thrust against a taking-off.

Note, to position yourself thus with the approach into the inverted thrust, stand with your left foot forward, your left hand in front by your left knee on your spear, your right between both of your legs back by your point, [with] your forward point toward his breast, simultaneously quickly step inside with your right leg, and there to be [as if] you want to thrust him to his right side, simultaneously invert your hands and thrust him to his left side with an inverted thrust.

If he then thrusts thus toward you, and you stand in your taking-off [with] your left foot set in front, [and] your spear forward on the earth up high before your face with extended arms, then take him aside below and above from one side to the other, in this, pass forward with your right leg, and wind yourself upward with your spear onto your right side, and thrust him to his throat. If he displaces you, then change through from your right onto his left side, and with this thrust him at his face. If the enemy notices the thrust, and then sets you aside, then pass forward with your left, and thrust him to his left side.

If he thrusts toward you thus, and you stand with your right foot forward, then wind your right hand onto your left side, then with this set his thrust aside, simultaneously withdraw your right leg to the back, and turn yourself with your spear again onto your right side on the strong, simultaneously pass forward again with your right leg, and thrust him to his upper opening, and withdraw yourself with this to the back into a good defense.

Cod. Vinob. 10825, 170r
A high charge-in against a high parry.

When you move toward an enemy by properly wielding the lance, place the left foot in front, grip the middle of the lance, the longer end should be tilted backward, the arms greatly extended on the lance, the left hand forward, [and] surely, if the right is positioned behind the head, you will strike inside both of the enemy's arms toward the chest with the forward point.

But if he advances toward you using this same technique, with the left foot forward, and he aims for your chest, [then] bring the left back, and parry the thrust between both your arms. However, if you also suddenly leap backward with the right, and quickly wind the spear through the hands, so that the forward point is in the left hand, [then] thrust the same [point] into either the enemy's eyes or his chest. But if he parries, and closes in on you while attacking, [then] you should pass backward while rotating the lance twice, so that you are holding its rear point with the right hand on the right side.

Hereafter, if the enemy employs the same thing against you, then you should thrust the lance extended into his groin, and following with the right foot, stab the adversary's chest. But if he parries this attack, [then] while extending the lance, if you rotate it from his left side toward the right, you should wind into the enemy's face. On the other hand, when the enemy parries this thrust, following with the left, attack his left side, and at the same time you will withdraw from the enemy while guarded.

C93, 198r
An upper running-in against an upper taking-off.

Note, when you come toward the opponent with the approach, then position yourself thus with your spear, [with] your left foot forward, and take your spear in the middle, allow your long point to hang to the back, your arms properly extended on your spear [with] the left hand forward, the right behind your head, simultaneously thrust him with your forward point in-between both of his arms to his breast.

If he runs at you thus, and you stand with your left foot forward, and if he thrusts you to your breast, then again withdraw your left leg to the back, and set his thrust aside in-between both of your arms with your spear, simultaneously also spring back quickly with your right foot, and let your spear shoot quickly through your hands thereby that you will have the forward point in your left hand, simultaneously thrust him with your forward point to his face or the breast. If he displaces you, and travels after you, then change yourself quickly twice to the back so that you again have your spear by your back point in your right hand on your right side.

If he changes himself from you to the back, then shoot him with the long [of] your spear to his privates [groin], in this, pass forward with your right leg, and thrust to his breast. If he sets you aside, and you have prepared your spear again into the long, then change it through from his left onto his right side to his face. If he notices the thrust, and then displaces you, then step with your left leg back, and thrust him to his left side, and withdraw yourself then twice to the back into a good defense.

Cod. Vinob. 10825, 170v
A low charge-in against a low parry

In the bind, you will position yourself against the enemy in this way: if you move in close, you should bind the forward part of his lance with your lance, the left foot leading. During this same bind, you will certainly watch whether he holds the lance strongly or weakly. If [his] hand strongly resists, then by quickly passing the lance backward through the hands, move toward him with the right foot, and aim for his throat with a thrust.

However, if he employs this same technique against you, the left foot set in front, and you place the left hand near the left knee, the right foot near the rear point, [then] you should pass toward the enemy with the aforementioned foot; and you will not hesitate to parry his own thrust to your right side, between your hands; at the same time, also bring the right foot back, and wind the lance onto the right side. In fact, if you pass the lance back during this same wind, [then] from this guard, you will be able to come to equal efforts in the struggle when the forward point is extended, and if you wind the lance around his lance at the same time, from the enemy's left side toward the right, [then] you will shoot a thrust into his eyes with vigor.

On the other hand, when the adversary employs a similar technique against you, you will remember to parry him toward your right side with the forward point, but you should also thrust the forward point of the lance from his right side toward the enemy's eyes, with the right foot brought back, and [then] you may withdraw from him in your protected guard.

C93, 198v
A lower run-in against a lower displacement

Note, position yourself thus with this approach to the opponent when you are to fight, then bind him forward on his spear so that you stand with your left foot forward, and in the bind feel whether he is weak or hard as perceived on his spear, if he is opposing you weakly, and if you hold strongly against [him], then let your spear quickly run through your hands to the back, simultaneously step inside with your right leg, and thrust him to his throat.

If he is runs-in at you thus, and thrusts you to your throat, and you are standing with your left foot forward, [with] your left hand by your left knee, [and] the back point by your right leg, then step inside with your right foot, and set his point aside in-between both of your hands with your spear on your right side, in this, set you right leg again to the back, and wind yourself with your spear onto your right side, and in this wind, let your spear run through your hands to the back, then you can come at him to the same work with your forward point, simultaneously change it through on his spear from his left side onto his right, and with this thrust him to his face.

If he changes you through thus, and thrusts you to your face, then set him aside with your forward point onto your right side, and again withdraw your right leg to the back, simultaneously shoot him [with] your spear onto his right side with your forward point to his face, and withdraw yourself with this to the back into a good defense.

Cod. Vinob. 10825, 171r
A parry against a free thrust which comes with all [his] strength, so that you may thrust the lance toward the enemy's eyes

You will position yourself in this stance using the following method: you will place the left foot in front, with the lance raised above the head, the right hand positioned at the rear point, while the left should be extended toward the middle of the lance, and the forward point is to have been tilted toward the ground. And if it should happen that, from this position, an adversary who is pointing the lance at your face notices whether you are resisting fiercely with the lance, and you are attacked [with] the left foot leading, [then] you should parry his attack on the right and left side, and by also changing the lance from your right side toward the enemy's left, you will lift the forward part of the lance up high, and following with the right, you should thrust the lance into the adversary's eyes.

However, if the same thing happens to you from him, [then] you will pass backward with the left, and if you invert the right hand on the lance, having rotated it onto your right side, his thrust will be blocked. In fact, using this technique from the bind, attack toward the adversary's right side [into] his eyes. Moreover, if he repels you by setting [your thrust] aside, then thrust the lance into the enemy's chest by winding from his right side toward the left.

However, in the event that he employs this same thing, while you are placing the right foot forward, [then] quickly thrust the lance into his groin by winding the lance from his left side toward the right, and from this position you should retreat backward, nevertheless taking care that it is done while guarded.

C93, 199r
A take-off against a free thrust thereby from the large hole

Note, position yourself thus in the take-off against a free-thrust, stand with your left foot forward, your spear well over your head, your right hand back by your point, your left will be forward on the spear so that the forward point lies on the earth. If he thrusts then into you from the strong, [with] his left foot set in front, then take him aside from your right and left side, and go up with the front of your spear, simultaneously pass forward with your right foot, and thrust him to his face.

If he thrusts you to your face thus, then step again with your left leg to the back, and exchange the right hand on your spear, and wind your spear onto your right side, then with this you will take his thrust aside, in this, go in from the bind to his face on his right side. If he sets you aside, then wind him from his right onto his left side to his breast.

If he thrusts you to your breast thus, and you stand with your right foot forward, then change through quickly from his left onto his right side, and with this thrust him to his privates, simultaneously withdraw yourself, in this, change to the back, and having taken good care of your face,

In this, wind him from your right onto his right side.

Cod. Vinob. 10825, 171v
A bind from which you may advance with a charge, by which the enemy is thrown.

When you want to use this technique at the right moment, you will place the left foot in front, and bind the lance not too strongly on the enemy's right side; however, by changing it through, quickly aim for the left side with a thrust.

However, if he attacks you in the same way, then, anticipating it, you will parry him with the middle of the lance; follow with the right, and if you lift the lance above your head with both hands, [then] thrust the lance into the adversary's chest or eyes with all your might.

On the other hand, when he attempts the same thing against you, with the lance thrown down, you will push his lance up away from your chest with the left hand, meanwhile, you will immediately advance to the inside, and seize the middle of the enemy's lance.

But if he grabs yours using the same technique, [then] you will remove the left hand from the lance, and seize the weak part of his right arm, near the elbow, and place the right foot behind both his feet. Meanwhile, definitely grab the weaker part of his left arm, on the back of the elbow, with the right hand, and if you arrange your arms in the shape of a cross with all your might, you should meanwhile be able to reach your head. Consequently, if you turn yourself around in this position, [then] whenever you want, it will be possible to carry the enemy on [your] back, without any damage to your body.

C93, 199v
A bind from which a run-in goes with a throw.

Note, position yourself in this bind thus, stand with your left foot forward, and bind him with the strong on the spear from his right side, simultaneously quickly change it through, and thrust his left side.

If he then thrusts you to your left side thus, and you shall notice [it at] the same time, then take it aside with your half-spear, pass forward with your right leg, and going with both hands with your spear well up over your head, then you will have to go on into a full thrust, and with this thrust him to his face or the breast.

If he thrusts from the strong at you thus, then let your spear fall, and wind his spear from below up from your breast with your left hand, simultaneously step inside quickly, and catch his spear in the middle.

If he guards your spear thus [with a] strong catch, then loose your left hand from your spear, and with this seize at his weak, behind the elbow of his right arm, step with your right leg behind both of his legs, in this, seize with your right hand onto his left arm on the weak, behind the elbow, and cross your arms with all of [your] strength that you with the head in-between both of your arms will come back through, and turn yourself with this, then [you will carry] him on the back where you on there will have [him] and he may not bring you any harm.

HALBERD

Cod. Vinob. 10825, 178r
The first two high strikes of the halberd from the left side.

You will compose yourself in this stance in this way: place the left foot forward, [and] hold the halberd above the head. Meanwhile, certainly following with the right foot, you will batter the enemy's head.

But if he attacks you with this same technique, [you] standing in the high strike guard, and [your] left leading, then bring the left backward, and if you should strike together with the adversary, his blow will be blocked; in fact, with the halberd raised up, you will simultaneously aim for the enemy's face with a thrust of the forward point. But if he parries, [then] by rotating it [the halberd] from his left side toward the right, press the blade of the halberd to his right arm.

But if he employs the same thing against you, [then] if you step back with the right foot, you will free yourself from the enemy by striking, but you will next thrust the halberd up into his vision, and during this same technique, you will bind your halberd [with his][1] on his left side, [and] from this position of the halberd, if you shift your weight to the rear, [then] you will pull toward yourself. From here, if you feel that the enemy is resisting strongly, [then] following again with the right, thrust into the adversary's chest.

On the other hand, with him having made use of this technique, with the left brought back, you will also parry with the rear point, and with this technique you will avenge yourself without danger from him.

C93, 202r
The first two upper-strikes of the halberd from the left side.

Note, you are positioned in this technique with the approach thus[2], stand with your left foot forward, and hold the halberd over your head, simultaneously pass forward with your right foot, and strike him to his head.

If he then strikes you to your head thus, and you also stand opposing him in the upper-strike with your left foot forward, thereby set it then to the back, and strike also from above at the same time as his, then is his strike in vain, in this, go up and thrust him to his face with your forward point. If he displaces you, then change through from his left side onto his right side, and set upon his right arm with your blade.

If he has you thus confined, then step with your right leg to the back, then you strike yourself from him with this, in this, strike up with your halberd in before his face, and in this strike, bind him on his halberd onto his left side, simultaneously turn your halberd, and pull with this to you. If he is strong, and you will also abandon this, then again step inside with your right leg, and thrust in to his breast.

If he thrusts you to your breast thus, then step again with your left leg to the back, and set his thrust aside with your back point, then you will be free from his harm.

---

1. This clause is full of ambiguities in the Latin and literally reads "you will bind your halberd with yours on his right side." Based on the German, *tuis* appears to be a misspelling of *suis*, "to/with his [halberd]."

2. The Vienna reads *Item schickn dich also mit disem zufechten* or "Note, position yourself thus with this approach."

Cod. Vinob. 10825, 178r
An under-strike with the halberd from both sides.

For the purposes of this under-strike, which will soon follow, you will position yourself in this way: place the left foot forward, the point should angle down toward the adversary, and immediately move the right hand to the right hip, and then thrust the halberd toward the enemy's eyes or chest. If he sets your thrust aside, then wind the halberd from his left onto [his] right side.

However, if he employs a double thrust against your face, with [your] left foot leading in the position of an under-strike, then while holding the halberd with the right hand, bring [it] to [your] right thigh, and you will stop the adversary's thrust with the blade on your right side; following with the right [foot], certainly thrust the forward point of the halberd into his chest.

But if the enemy employs the same thing against you, [then] you will bring the left foot back, and having next brought the right back in a similar fashion, you will come to equal efforts in the battle with the enemy. But also immediately break his head open with a strike from above.

However, if he attacks you from above, having quickly raised the halberd up high, [then] you will parry his blow with the blade of your halberd, and if you attack his groin with a thrust, it will be possible to retreat backward.

C93, 202v
The lower-strike of the halberd from both sides.

Note, position yourself thus with the approach into the lower-strike, stand with your left foot forward, and hold the point down toward the opponent, in this, go with your right hand on your right hip, and with this thrust him to his face or the breast. If he sets your thrust aside, then change through from his left onto his right side.

If he thrusts twice to your face thus, and you also stand with your left foot forward in the lower-strike, then go with your right hand on your right hip with your halberd, and set him aside onto your right side with your blade, simultaneously pass forward with your right leg, and set your forward point upon his breast.

If he sets the point onto your breast thus, then step with your left leg to the back, simultaneously also quickly withdraw the right to the back, then you may again come to the same work with it, and strike him with an upper-strike to his head.

If he strikes you from above thus, then go up with your halberd from below, then you will catch his strike on your blade, simultaneously pull back, and thrust him to his groin, and withdraw with this to the back.

Cod. Vinob. 10825, 179r
The crossed changer against a position of setting aside.

When you want to make proper use of the aforementioned technique, you will place the left foot forward, the left hand should be placed on the haft underneath the right [arm, as depicted by the left figure], and from this position, step to the inside with the right, wind twice into an under-strike, and smash the adversary's left foot with a thrust.

But if you anticipate the same thing to be done against you by the enemy, standing in a position of setting aside and the left foot leading, then if you bring the right hand up, his thrust will be rendered useless, and immediately following with the right foot, you will thrust the forward point into the enemy's face.

However, when he attacks your face, you will not hesitate to set him aside with the strong part, from your left side to the right; in that same instant, strike the halberd from both sides into the enemy's eyes. Moreover, from this same strike, if you bind his halberd, [then] having rotated it twice, attack his eyes with a thrust.

However, if he uses the same thing against you, [then] parry with the blade of your halberd in such a way that his thrust is deflected above your head. Furthermore, from this technique, follow with the left foot, and if you push the adversary's halberd away with the rear point, you will make yourself free and guarded.

C93, 203r
The crossed changer against a take-off.

Note, position yourself thus in the crossed changer, stand with your left foot forward, your left hand under your right shoulder, simultaneously step inside with your right leg, wind yourself with this twice into the lower-strike, and thrust him to his left leg.

If you notice this thrust, and stand against him in the take-off, [with] your left foot set forward, then go with your right hand onto your right side, then you will take away his thrust, simultaneously pass forward with your right leg, and thrust him [with] your forward point to his face.

If he then thrusts you to your face thus, then set him aside from your left onto your right side with the middle of the strong, simultaneously strike up from one side to the other into the front of his face, and in this, from the strike bind him on his halberd, in this, change him through twice, and thrust him to his face.

If he thrusts toward you thus, then take him aside with your blade so that the thrust goes out over your head, in this, pass forward with your left leg, and strike his halberd away with your back point, then you are free from him.

Cod. Vinob. 10825, 179v
A thrust that aims for the eyes against [a strike] which is formed from the chest, and which also cuts his chest.

For this technique, you will compose yourself in this manner: set the left foot in front, hold the halberd near the chest in such a way that the blade is turned upward, and following with the right [foot], attack the adversary's eyes.

However, if the adversary uses the same thing against you, the left foot leading, [then] you should direct the strike which is formed from the chest up toward him, and set his thrust aside in this manner. Quickly following with the right foot, you will by all means strike his chest with a thrust. But if the enemy sets it aside by parrying, then you should wind your halberd around his, from his right side toward the left; from that point, if you also place the left foot in front of his right, [then] attack the enemy's eyes with a double thrust.

However, when he uses the same thing against you, you will defend yourself by parrying his attack on both sides. You may [then] advance to the inside with the left foot while rotating the halberd from the adversary's right side to the left, [and] smash his left arm with a thrust. When the enemy parries this attack, then you will guide the rear point of the halberd toward the enemy's chest by raising up away from the ground.

But if he uses the aforementioned technique against you, [then] step back with the left foot [and] his change-through will be hindered; meanwhile, you will immediately stab his face with a thrust, and withdraw backward.

C93, 203v
A face-thrust against the breast-strike.

Note, position yourself into the face thrust thus, stand with your left foot forward, and hold your halberd at the breast so that your edge is turned up, simultaneously pass forward with your right leg, and thrust him to his face.

If he thrusts you to your face thus, and you stand with your left foot forward, then go up with your breast-strike, and in this, set his thrust aside, then step inside with your right leg, and thrust him to his face.

If he thrusts twice at you thus, then set his thrust aside from both sides, and step inside with your left leg, and change him through from his right onto his left side, and with this thrust him to his left arm. If he displaces your thrust, then wind him [with] your back point from below in to his breast.

If he winds you [with] his back point to your breast thus, then step with your left leg to the back, then is his wind in vain, in this, thrust him to his face, and withdraw yourself with this to the back with the full and half halberd.

Cod. Vinob. 10825, 180r
A bind or clash of the halberds from which follows a pulling technique.

For the aforementioned method, you will adjust yourself thusly: place the left foot in front, the rear point of the halberd will have been brought to the right hip, the forward [point] pointed at the adversary's eyes, and from your right side toward the enemy's left, bind his halberd, and at that moment you will quickly curve your halberd above his, and from this position, pull toward you.

However, if you are standing thus against the enemy in the position of a bind of the halberds, left foot leading, and he pulls toward himself, then follow with the right, and thrust above his arm toward the chest. But if the adversary anticipates the thrust, and also attempts to set [you] aside, then by rotating or winding the halberd from the enemy's left side to his right, drive the halberd toward the adversary's eyes.

But if he shifts from one of your openings to another, [then] bring the left foot back, and [parry] his attempt on both sides with the rear point. When this has thus been executed, you should pass to the inside again with the left foot, and with the blade of the halberd extended, you should parry his halberd from this guard; also, quickly thrust it into the enemy's eyes. However, if he parries it, [then] if you seek out one opening from another while changing, remember to withdraw twice from the adversary.

C93, 204r
A bind from which goes a pull.

Note, position yourself thus in the bind, stand with your left foot forward, your halberd with your back point on your right, the point toward his face, and bind him on in front of your right side onto his left, in this, wind your halberd quickly over his, and pull with this toward you.

If you also then stand against him thus in the bind with your left foot forward, and he pulls to himself thus, then pass forward with your right leg, and thrust him to his breast inside over his left arm. If the opponent notices this thrust, and then displaces you, then change through from his left onto his right side, and thrust him to his face.

If he seeks your openings from one side to the other, then step with your left foot to the back, and set him aside with your back point on both sides, in this, again step inside with your left leg, and wind yourself with the blade leading, and with this take his halberd away, simultaneously thrust him to his face. If he sets you aside, then seek his openings from one side to the other, in this, change through, and withdraw away with this twice.

Cod. Vinob. 10825, 180v
A bind of the halberds with an additional technique by which the adversary can be blocked above.

When you want to properly position yourself in this manner during a bind, you will remember to place the left foot forward. And if he in turn now stands against you in the aforementioned stance, [his] left foot leading in the same way, [then] bind the forward part of the enemy's halberd. However, during this same bind, you will wind the blade of your halberd over his curved point, which is connected to the halberd's blade.

But if he uses the same technique, [then] reverse the halberd, and with it held up high, quickly draw it to you from above, and pierce the adversary's eyes or chest. But if the adversary parries this thrust, then you will vigorously wave the halberd in his vision, and meanwhile seek out his nearest opening.

But if he in turn searches for your [openings] in the same way, then you should be able to parry him with the rear point, [and] immediately follow with the right; if you set his halberd aside with the back point, [then] by winding twice from the lower part upward, thrust the halberd toward the adversary's face of chest. However, if he anticipates this, and withdraws backward, [then] follow the enemy by pressing hard twice with strikes and windings.

But if he uses the same thing, then you will parry him with the forward point of your halberd together with the rear; from that point, you will withdraw backward while keeping a careful watch.

C93, 204v
A bind with an upper block.

Note, position yourself with this approach thus, stand with your left foot forward, if he then stands also against you in the same bind [with] his left foot set in front, then bind him on the front of his halberd, and in this bind, then wind your blade over his hook.

If he desires to wind you thus, then turn your halberd, and then shoot over it, and in this, pull your halberd up quickly, and thrust him to his face or the breast. If he sets your thrust aside, then strike him with your halberd up before his face, in this, you will seek his next opening.

If he seeks your opening thus, then set him aside with your halberd's forward point, pass forward with your right foot, and strike his halberd away with your back point, and wind him with this twice from below to his face or the breast. If the opponent notices the wind, and with this steps to the back, then travel after him twice with striking and with winding.

If he travels after you thus with this technique, then set [him aside] thus with your back and forward point of your halberd, step with this to the back, and have your face also with a good defense.

Cod. Vinob. 10825, 181r
A technique that is able to strike the enemy from above, after which is added a pulling technique from below.

In regard to this technique, you will position yourself thusly: with the left foot set in front, you should hold your halberd in a high strike guard. And if he in turn places his left [foot] against you, with his halberd extended toward your face, then you should repel him with a high strike. Meanwhile, you will also pull back toward yourself from your left side to the right; from this position, you should consequently direct the halberd toward the enemy's eyes.

But if he employs the same thing, [then] you will hook the blade of your halberd behind his foot, and if you draw it toward you, his attempt will be made ineffective, and from this position, you will also be able to throw the enemy. Furthermore, immediately following by pressing forward, you will smash his eyes with a thrust.

However, if he pursues you in the same manner, [then] you will parry him with the middle of the halberd, and while rotating it from one side to the other, you will seek out the enemy's nearest opening. In the event that he in fact frees himself from this dilemma, you should strike him at that exact same moment. But also, during this same halberd bind, you will parry his front [point] with the rear point; you will place the right foot on the inside of the enemy's left; meanwhile, certainly hook the rear point around his neck from the right side, and if you pull the adversary from this position, you will easily throw him down.

However, if he attempts to throw you down, you will switch both hands, and apply your halberd to his right arm; from this position, if you push him off, you will free yourself.

C93, 205r
An upper inside strike with a lower pull.

Note, position yourself thus in the upper inside-strike, stand with your left foot forward and hold your halberd in your upper-strike, if he then also stands against you [with] his left foot forward, setting his halberd toward your face, then take him aside with your upper-strike, then simultaneously pull to you from your left on to your right side, and in this pull, go up with your halberd in to his face.

If he wants to go to your face thus, then turn your halberd with the blade behind his leg, and with this pull it back to yourself, then is his thrust in vain, and he may also fall with this, in this, travel after him with a thrust to his face.

If he thrusts you to your face thus, then take him aside with your half halberd, and change through from one side to the other, and with this seek his next opening. If he sets you aside, then bind equally with him on his halberd, and in this bind, wind his forward [point] away with your back point, and step with your right foot in quickly for his left, simultaneously drive in with your back point around his throat on his right side, pull in with this toward you, then [you] should [go] into a throw.

If he want to throw you thus, then exchange both hands [on] your halberd, and set your halberd onto his right arm, shoot in with this from you, then you shall be free from his harm.

Cod. Vinob. 10825, 181v
A method of winding above, by which the enemy's halberd is forced down after the initial bind, with an additional lower block.

For the purposes of this technique, you will compose yourself in this manner: place the left foot forward; you should hold the halberd above the chest on the right side; the point is to be aimed at the adversary's eyes; from this position, you may step to the inside with the right, and using an inverted strike, the strike that is upside-down, if you hold the halberd near the front part, and the right hand is certainly wound around the rear, below the left arm, so that you may beat his head with the forward part, [then] smash his head.

But if he attacks you from above, the left foot leading, then parry him with the halberd's blade, and you should strike at the same time as the enemy. Furthermore, during this same halberd-bind, by rotating it from his left side to the right, over the enemy's halberd, change yours over, and if you push down strongly, the adversary will be blocked.

But when he employs the same thing, then following with the left, you will strongly lift your halberd up high by changing, still from this position, so that the point may be extended toward the adversary's eyes. When the enemy parries it, you will place the right foot in front of his left, and you will remember to strike the rear point between both of his arms [and] over the right arm. From this position, you will consequently push down from your right side; you will be able to strike his head with the blade of your halberd.

However, if the enemy attempts to do the same thing to you, then you will switch the left hand, gripping the halberd backward; if you bring the left foot back, stab the enemy's head with the rear point.

C93, 205v
An over-wind from the bind with a lower block.

Note position yourself in this technique thus, stand with your left foot forward, your halberd at your right breast, your point toward his face, simultaneously step inside with your right foot, and strike him with a inverted strike to his head.

If he strikes you [from] above thus, and you stand with your left foot forward, then displace him with your blade, in this, bind equally with him, and in this bind, change through from his left onto his right side, over wind his halberd with this, and press his strongly down.

If he guards you thus [with a] block, then pass forward with your left foot, and wind yourself with your halberd again strongly up [with] the point in to his face. If he sets you aside, then step with your right foot for his left, and wind yourself with your back point in-between the inside of both of his arms over his right arm, press him down with this onto your right side, then you will want to [go] in with your blade to his head [with] a strike.

If he has wound you hard thus, then quickly again change your left hand through on your halberd, simultaneously step with your left foot to the back, and strike him with your back point to his head.

Cod. Vinob. 10825, 182r
A thrust to the inside, by which the arm is attacked, against a thrust to the eyes.

If you move toward one another, and the adversary attacks your eyes with a thrust, [then] you should step to the inside with the left foot, and smash his left arm on the inside with a thrust, and having immediately connected the halberd's blade to the same [arm], you will in this manner turn the enemy toward his right side, and his thrust will be reduced to nothing; from that point, you will be sure to draw the halberd back, and [then] attack the enemy's chest with the forward point.

But if, in fact, he should threaten your chest in this same way, [then] you would not hesitate to bring the left foot back, and you will parry him with the blade—that is, with the [. . .] part of the halberd; afterward, you will raise the halberd into his vision, and stepping to the inside again with the left, you will thrust the point into either his face or chest.

But if the enemy employs the same thing, [then] you will set him aside by defending yourself from one side to the other, and from this position, if you advance to the inside with the right foot, [then] you will beat his head with the back point. When the adversary parries it, then if you pass toward the enemy's right side with the left foot, you will smash his right arm with your halberd's blade.

However, if he employs the very same thing against you, [then] you should parry him with the rear point, and thrust the front [point] underneath, toward his chest. Afterward, following with the right foot, beat the adversary's head with the rear point.

C93, 206r
An inside arm thrust against a face thrust.

Note, when you come together with the approach, and he thrusts you to the face, then step inside with your left foot, and thrust him inside quickly to his left arm, and in this, set the blade on him, then you will shoot in with this from you onto his right side, and then is his thrust in vain, simultaneously withdraw your halberd quickly toward you, and thrust him with your forward point to his breast.

If he thrusts you to your breast thus, then set your left leg to the back, and set him aside with your blade, in this, go up with your halberd for his face, step inside again with your left foot, and wind your point to his face or the breast.

If he winds you to your face thus, then take him aside from one side to the other, in this, step inside with your right foot, and strike him with your back point to his head. If he displaces your strike, then step with your left foot by his right side, and strike him with your blade to his right arm.

If he strikes toward you thus, then take him aside with your back point, and wind your forward [point] from below to his breast, simultaneously step inside with your right leg, and strike him with your back point to his head.

Cod. Vinob. 10825, 182v
A thrust to the chest, with an additional simultaneous block.

When you want to properly employ the aforementioned technique in an athletic and vigorous manner, you should place the left foot forward, hold the halberd above the right part of [your] chest, and attack the adversary's chest with a thrust using the front point.

But on the other hand, if you take position against the enemy with the right foot leading, and he attacks you in the same manner, then you will parry him with the forward part of your halberd; you will also immediately guide your halberd above the blade of the adversary's halberd, and from this position, if you turn yourself to your left side, you will block his halberd. During this same block, definitely move the left hand closer to the right, and if you bring the right foot back, you may strike toward his head.

However, if he uses the same thing against you, [then] you should bring the left foot back, and parry his own strike with the front point; at that moment, you will also strongly lift it up, pass to the inside with the right foot, and apply the rear point of your halberd to his neck. Moreover, if the enemy parries this, [then] quickly following with the left, smash the halberd's blade into his head.

But if the adversary uses this exact same technique, then parry his attack between both hands on your right side, and by striking [him] twice, you may withdraw.

C93, 206v
A breast thrust with a block.

Note, position yourself into this technique thus, stand with your left foot forward, and hold your halberd over your right breast, simultaneously thrust him with your halberd's forward point to his breast.

If you then stand also opposing him with your right foot forward, and he thrusts you thus, then set his thrust aside in front with your halberd, simultaneously attack him with your halberd over his blade, and invert yourself with this onto your left side, then you will block his halberd, and in this block, seize with your left hand to your right, in this, set your right foot to the back, and strike above to his head.

If he strikes you above thus, then step with your left leg to the back, and set him aside with your halberd's forward point, in this, shoot in strongly with your halberd above his, step inside with your right foot, and wind your back point to his throat. If he sets you aside, then pass forward with your left foot, and strike him with your blade to his head.

If he strikes you above thus, then displace him in-between both of your hands on the halberd onto your right side, with this you will strike twice to withdraw.

Cod. Vinob. 10825, 183r
The lower strike to the chest against a high strike, together with an additional wrenching technique.

In regard to the halberd-clash, you will position yourself by wielding [the weapon] in the following way: you will place the right foot in front, direct a lower chest-strike upward toward his left arm, and by switching [your] hands on the halberd, if you turn yourself from the right side to the left side, then pull toward you.

But if the adversary attempts the same thing against you, the left foot leading, [then] you will immediately change the left hand on your halberd, and strike his head from above. When the enemy parries this strike with the forward point, then you will wind up toward his chest on his left side.

If you anticipate a thrust from the adversary, [then] with the right foot brought back, you will parry him with the rear point. Meanwhile, you will also work from one side to the other with this same point, follow with the right, remember to strike your halberd between both of his arms toward the enemy's chest, and from this position, curve your halberd above his left arm. But if the enemy's hand is not removed from his halberd, [then] you will break his arm.

However, when he attempts to break your arm with the same technique, you will move your hand, and following to the inside with the right foot, you will certainly strike the enemy's groin with a thrust using the rear point; you will free yourself from him without any trouble.

C93, 207r
A lower breast-strike against an upper-strike with a wrench.

Note, position yourself in the approach thus, stand with your right foot forward, and go up with the lower breast-strike into his left arm, and exchange your hand on your halberd, in this, wind yourself from your right onto your left side, and pull with this to yourself.

If he strongly pulls you toward him thus, and you stand with your left foot set in front, then change your left hand quickly through on your halberd, and strike [from] above to his head. If he displaces you with his forward point, then wind from below to his breast on his right side.

If you notice the enemy's thrust, then set your right foot to the back, and take [him] aside with your back point, simultaneously work with the same point from one side to the other, pass forward with your right leg, and wind your halberd in-between the inside of both of his arms to his breast, and turn with this your halberd over his left arm. If he will not release his hand from his halberd, then you will break his arm.

If he wants to go at you to break the arm thus, then exchange your hand on your halberd, simultaneously step inside with your right foot, and thrust him with your back point to his groin, then with this you will become free from him.

Cod. Vinob. 10825, 183v
Two pulling techniques from above, following the first bind or clash of the halberds.

When you engage each other by wielding the halberd athletically, and the halberds are both bound, you will place the right foot forward, and hook the front blade around the adversary's neck.

But if he attacks you with the blade of his halberd in the same manner, the left foot leading, [then] you will in turn hook your halberd around his neck, and during this same wind, pull it strongly toward you with all your might.

However, if the enemy makes use of the same technique, you should pass toward him on the inside with the left foot, and push his halberd off to your left side with the rear point; also, quickly apply that same point to the front part of [his] neck, place the right foot behind his left, and if you pull toward your right side, you will throw the adversary to the ground.

On the other hand, if he sets up to do the same thing against you, [then] remove your left hand from the halberd, and at the same time, if you seize [his] right arm above the elbow, and push the enemy, you will free yourself. Without delay, you will definitely grip the halberd again with the left, strike his head, and by changing, you may withdraw from the enemy.

C93, 207v
Two upper pulls from the bind.

Note, when you both come together with the approach, and have one other bound, then step with your right foot before his left, and wind him [with] your blade forward around his throat.

If he wants to pull you with his blade thus, and you stand with your left foot set in front, then wind him also with your halberd around his throat, and in this wind, then pull strongly toward yourself.

If he then pulls strongly toward himself thus, then step inside with your left leg, and take his halberd away with your back point onto his left side, in this, drive in with your back point forward around his neck, and set your right leg behind his left, [then] pull with this onto your right side, then you will want [to go] into a throw.

If he wants to throw you thus, then release your left hand from your halberd, and seize with this well behind his right elbow, push with this from you, then you will be free from him, simultaneously quickly seize with your left hand again to your halberd, and strike him to his head, in this you [will] change, to withdraw from him with this.

Cod. Vinob. 10825, 184r
A lower pulling technique against a violent thrust.

You should compose yourself in this manner: place the left foot forward; you should hold the halberd near the left foot, the blade facing the adversary. But if the enemy takes position against you for a violent thrust, leading with the left foot, and he attempts to stab you, then hook your halberd around his more forward foot; in this way, if you shift your weight from your right side toward the left, you will pull the enemy in such a way that his thrust cannot hurt you, and at the same time you will be able to throw him.

On the other hand, if he attempts the same thing against you, then suddenly remove your halberd from his neck, be sure to hook your halberd's blade around his halberd, and if you push down strongly, you will free yourself, so that you will not be thrown by the enemy. Meanwhile, you should certainly lift your halberd up away from his, into his eyes. Otherwise, if the enemy removes it by defending himself, [then] while rotating the halberd from his left side to your right, you will crush his throat with a thrust. But if he in turn parries it, [then] step to the inside with the right, and curve the rear point toward the adversary's eyes.

However, if he should use the same thing, you will not hesitate to bring the left foot back, and parry his forward point with the back of your halberd; meanwhile, you should also pass to the inside with the left, [and] if you direct the front point toward the enemy's eyes, you may may withdraw from the adversary in a changing guard.

C93, 208r
A lower pull against a powerful thrust.

Note, position yourself thus, stand with your left foot forward, and hold your halberd properly by your left leg, the blade toward the opponent. If he then stands against you in the face thrust, [with] his left foot set in front, and wants to thrust you from this, then go in with your halberd around the front of his leg, and turn yourself with this from your left onto your right side, then you will pull in to yourself, then his thrust cannot bring you any harm, and you may also go into a throw.

If he prepares to guard you thus, and desires to throw you, then loose your halberd quickly from his throat, and simultaneously set the blade upon his halberd, and press his strongly down, then you are free of the throw, in this, drive in on his halberd quickly up to his face. If he displaces you, then change through from his left onto his right side, and thrust him to his throat. If he goes to displace you, then step inside with your right foot, and wind your back point to his face.

If he works toward you thus, then set your left foot to the back, and take his forward point away with your back, simultaneously step inside with your left foot, and wind him [with] your forward point to his face, with this, withdraw yourself to the back into the changer

Cod. Vinob. 10825, 184v
A wind with a simultaneous method of throwing the adversary.

When you meet by wielding the halberds athletically, you both will bind halberds in the same moment; you will wind your [weapon] around his halberd from the adversary's right side toward the left, and beat his arm with a thrust. However, if the enemy anticipates this, and parries your attempt, you will wind the halberd around twice, and find his nearest openings.

But if he probes your [openings] in the same way, then you will hurl your halberd into his eyes, but also during this same hurling technique, you will strike the enemy's leading arm, and firmly strike away his halberd; consequently, if you wind around his [weapon] as quickly as possible, aim for his eyes with a double thrust.

But if he attacks you with the same technique, [then] parry with the front point of your halberd, the left leading.

On the other hand, if you in turn take position against the adversary, the left leading in this same way, then push aside his halberd with the rear point; you should place [your] right foot behind the enemy's left, and hook the rear point around his neck [as depicted by the right figure], and if you pull [his] upper [body] strongly toward you, you will throw the enemy using your right foot.

C93, 208v
An inside wind with a throw.

Note, when you both come together with the approach, and have equally bound with one another, then change through on your halberd from his right onto his left side, and thrust him with this to his left arm. If he notices the thrust, and displaces you, then change through twice, and seek his next opening.

If he seeks your next opening thus, then strike with your halberd from below on in before his face, and from in this blow, strike him to his forward arm, simultaneously bind him strongly on his halberd, and quickly change through, and thrust him twice to his face.

If he thrusts toward you twice thus, then take him aside with your halberd's forward point so that you stand with your left foot forward. If you then also stand against him with your left foot forward, then take away his halberd with your back point, step with your right foot behind his left, and set your back point up front around his throat, pull up with this strongly to yourself, then you will throw [him] in over your right leg.

Cod. Vinob. 10825, 185r
An upper transition used with a thrust to the side, as illustrated below.

Of course, by properly wielding the halberd in the aforementioned manner, you will position yourself in this technique, which will soon be described; you will place the left foot forward; you will hold your halberd on the right side; the point is to be extended toward the adversary. And if he in turn stands against you in the same position, [then] you should strike [your weapon] at the enemy at once, and during the bind of the halberds, you will hook your blade around the curved point of the adversary's halberd, and from this position lift upward.

On the other hand, if he hooks your halberd in the same way, and he uses the same technique against you, then you will place [your] right foot behind the enemy's left. Be sure to immediately pull your halberd back up above, and thrust the rear point toward his left side from below. Furthermore, smash your blade into the adversary's head while bringing the right foot back.

But if he tries the same thing against you, [then] you will parry his attempt with the rear point; meanwhile, also following with the right, you will stab the enemy's left arm. Hereafter, if the adversary parries it, then you should strike the rear point between both of his arms toward the chest; from that point, certainly pass backward with the right foot, and if you batter the adversary's head, you will withdraw from the enemy while striking.

C93, 209r
An upper passover together with a side thrust.

Note, position yourself thus with your approach, stand with your left foot forward, and hold your halberd on your right side, the point toward the opponent. If he stands then also against you in the approach, then bind on him quickly, and in this bind, wind your blade behind his hook, and shoot over it with this.

If he prepares to guard your halberd thus, and shoots you over it, then step with your right foot behind his left, in this, pull your halberd up, and thrust him [from] below with your back point to his left side, simultaneously set your right foot again to the back, and strike him with your blade to his head.

If he works at you thus, then take him aside with your back point, simultaneously pass forward with your right foot, and strike him to his left arm. If he displaces you, then wind him with your back point in-between the inside of both his arms to his breast, simultaneously step with your right leg to the back, and strike toward his head, then you will remove yourself from him with this strike.

Cod. Vinob. 10825, 185v
A lower strike with a parry using the middle of the halberd.

You will position yourself against the adversary by wielding the halberd athletically in this manner: place the left foot in front; you should hold your halberd on the right side. But if the adversary takes position against you in the same way, in a lower striking guard, the right foot leading, and holding his halberd on the left side, then you will shoot the halberd toward the enemy's face.

However, if he uses the same shooting technique against you, smash his left arm by striking upward.

On the other hand, when he attacks your left arm, you will then deflect his attempt toward your right side with the middle of the halberd, and from this position, hook the blade of your halberd around his left arm from above, [and] press down.

But if he indeed presses down strongly, place [your] left foot behind the enemy's right, [and] smash the rear point into his head; however, if you quickly bring the right foot back, thrust the forward point into the adversary's chest.

Moreover, if he uses the same thing, then you should parry him with the middle of the halberd, and wind the rear point from underneath up over the enemy's left arm, and if you pull toward your right side, you will withdraw from the adversary in a high striking guard.

C93, 209v
A lower-strike with a displacement at the half halberd.

Note, position yourself thus against the opponent in the approach, stand with your left foot forward, and hold your halberd on your right side. If he then also stands against you in the lower-strike, [with] his right foot set in front, the halberd on his left side, then strike with your halberd in to his face.

If he goes to your face thus, then strike him from below to his left arm.

If he strikes at your left arm thus, then set him aside with your half halberd onto your right side, and in this, set your blade onto his left arm, and with this press it down.

If he then strongly presses down thus, then step with your left leg behind his right, and strike him with your back point to his breast.

If he winds you thus, then set him aside with your half halberd, and wind him [with] your back point [from] below over his left arm, pull with this onto your right side, and withdraw yourself with a upper-strike from him to the back.

Cod. Vinob. 10825, 186r
A cut, with an added technique for throwing the enemy, following a method of kicking the foot into the adversary's knee.

In this form, which will soon be described below, you will position yourself in this way from a high cut, place the right foot forward, [and] hold the halberd above [your] head, with its blade turned upward.

And if you take position against the adversary with the right foot forward, and you hold the halberd on your right side, [pointing] toward the enemy's face, you should step to the inside with the left, [and] apply the leading blade to his neck.

However, if the enemy uses the same technique against you, then force the rear point of your halberd below the enemy's left armpit; push toward your right side; if you bring the right foot back, you will cut the enemy's head open with your blade from above.

But if he attempts to cut you, [then] by raising the halberd up above [your] head, you should parry his cut with your halberd's blade; from there, you will place your right foot behind the enemy's left foot, and push his halberd back from above with the rear point.

However, when he fends you off in the same manner, [then] while bringing the left foot back, curve the rear point toward the adversary's left arm.

When the enemy uses the same technique, you will place the left behind his right foot, and apply the leading blade to the enemy's neck; smash his leading right foot on the outside with the halberd's curved spike; meanwhile, from this position you will be sure to push strongly forward above; and if you pull him toward you from underneath, the adversary will be knocked down.

C93, 210r
A cut with a throw from the heel-kick.

Note, hold yourself with the approach from the upper cut thus, stand with your right foot forward, and hold your halberd over your head so that the blade is up.

If you then also stand against him thus, [with] your right foot set in front, then hold your halberd on your right side in at his face, in this, step inside with your left leg, and set your blade on the front of his throat.

If he has you engaged thus, then set your back point under his left shoulder [armpit], shoot with this onto your right side, simultaneously step with your right foot to the back, and cut him with your blade up to his head.

If he cuts you to your head thus, then go up with your halberd well by your head, and displace his cut with your blade, in this, step with your right foot behind his left, and take his halberd up with your back point [and] away.

If he takes away your halberd thus, then step with your left foot to the back, and wind him with your back point to his left arm.

If he winds toward you thus, then step with your left foot behind his right, and set your blade forward on his throat, and strike him [with] the hook out quickly around his right foot, simultaneously push over strongly from you, and pull under to you, then he must fall.

Cod. Vinob. 10825, 186v
A technique by which the adversary is turned around, against an under-strike.

For the purposes of this technique, you will position yourself thusly: you will turn yourself around with the right foot on your left side in a high strike guard, in such a way that your left foot is turned toward the adversary, and from this position you should strike his head with a high strike.

However, if the enemy uses the same thing against you, the left foot leading, and the back point of your halberd tilted toward the ground opposite the enemy, then you will raise [the weapon] up high, and parry his strike with the same point. Then, advancing to the inside with the right foot, you will break the adversary's left arm by hitting with an under-strike.

But on the other hand, if he parries in the same way, you should bring the rear point of your halberd to your right side, and if you pass backward with the right foot, you will rotate the front point from his left side toward the right and into the enemy's eyes.

However, when he uses a similar technique, push his attempt aside with the front part of the halberd; in fact, if you wind yours around his halberd from one side to the other, seek out his openings. But if the adversary pulls it away, you should pass to the inside with the right foot, and displace his halberd with your halberd's rear point, and once you thrust [the weapon] into his face or chest, you may then withdraw from the enemy with a double-step in a change guard.

C93, 210v
An inverter against a lower-strike.

Note, position yourself into the inverter thus, turn yourself onto your right foot from your left side around into the upper-strike so that your left foot is turned toward the opponent, in this, strike him from the upper-strike to his head.

If he strikes above at you thus, and you stand with your left foot forward, your back point forward on the earth toward the opponent, then go on with the same point, and take him away with this strike, in this, step inside with your right leg, and strike him from the lower-strike from below to his left arm.

If he displaces you thus, then go with your back point onto your right side, simultaneously step with your right leg to the back, and change your forward point from his left onto his right side to his face.

If he changes at your face thus, then displace him forward on your halberd, in this, wind through on his halberd from one side to the other, and seek his next opening. If he sets you aside, then step inside with your right leg, and take away his halberd with your back point, thrust him with this to his face or the breast, simultaneously step, in this, change twice to withdraw.

Cod. Vinob. 10825, 187r
A technique by which the adversary's halberd is taken away, with an additional throw.

If you engage each other by wielding the halberd athletically, and you both bind the halberds together, then you will wind yours around the enemy's halberd from his right side toward the left, and also include a thrust. However, if the adversary parries you, [then] if you pass to the inside with the right foot, you will turn the broader part of his halberd to your right side with the rear point.

Furthermore, if he attempts to execute the same thing, [then] while bringing the left back, thrust the forward point into the enemy's chest. But if the adversary parries, then you should advance toward him with a double-step, and knock the rear point into his head.

On the other hand, if he uses the same [technique] against you, you should turn him aside with the middle of your halberd. Meanwhile, if you also seize the rear point of the enemy's halberd with the left hand, [and] the forward [point] with the right, in such a way that you hold both halberds in your hands, then if you put the left foot on his right knee, it will be possible to break his foot.[1]

On the other hand, if he seizes both halberds, and wants to use the aforementioned technique in this way, [then] you will remove the right hand from your halberd, and reach around the outside of his knee, [then] firmly push the enemy with your left hand on top; in fact, if you lift up from below, you will force him to fall down, and all his efforts will be in vain.

C93, 211r
A halberd-taking with a throw.

Note, when you both come together with the approach, and have equally bound with one another, then change through from his right onto his left side with a thrust. If he takes you aside, then step inside with your right leg, and take away his blade with your back point onto your right side.

If he wants to take away your halberd thus, then step with your left foot to the back, and thrust your back point to his breast. If he displaces you, then again step inside twice toward him, and strike him with your back point to his head.

If he strikes at your head thus, then displace him in the middle on your halberd, simultaneously seize his back point with your left hand, and with the right to his forward [point], thus you will have both halberds in your hands, in this, step with your left foot onto his right knee, then you will want to break his leg asunder [in two].

If he has both halberds held together thus, and desires to step upon you, then loose your right hand from your halberd, and go onto him quickly around his knee-bend, simultaneously shoot up with your left hand strongly from you and pull him [from] below almost up, then he must fall, and you can harm his work.

---

1. The illustration and physical mechanics of this technique suggest that *pedem* refers to the knee or leg, not the foot.

Cod. Vinob. 10825, 187v
A technique by which the adversary's foot can be broken, against a thrust which attacks the neck.

When it happens that you both advance toward each other in an athletic manner, and you both bind halberds at the same time, [then] you should hold your halberd toward the enemy's eyes, and with your halberd rotated around his, attack his eyes with a thrust. But if he anticipates your thrust, and sets [it] aside, [then] you should proceed to the inside with the right foot, and rotate the back point toward his left arm from the area below.

However, if he employs the same thing against you, [then] you will quickly switch the left hand on your halberd, and batter his chest with the forward point. But if he wards it off, strike the rear point into the enemy's head.

But if he attacks you from above with the same technique, [then] you should step backward with the right foot, and parry his own strike with the forward point; meanwhile, definitely bind the lower part of his halberd strongly, and if you lift up from this position, [then] you should proceed to the inside with a double-step toward him, drop [your] halberd, reach behind his left foot with the right hand, and forcefully push his knee[1] with the left hand, and in this way it will be possible to break his leg,[2] or to throw the enemy down.

If he in turn closes with you, [then] press your halberd to his neck, and if you push down hard, you will neutralize all his efforts.

C93, 211v
A leg-break against a neck-thrust.

Note, when you come together with the approach, and have bound immediately with one another, then hold your halberd toward his face, in this, change off on his halberd, and with this, thrust to his face. If he notices the thrust, and displaces you, then step inside with your right foot, and wind your back point from below to his left arm.

If he winds at you [from] below thus, then exchange your left hand quickly on your halberd, and thrust him with your forward point to his breast. If he displaces you, then strike your back point to his head.

If he strikes above at you thus, then step with your right foot to the back, and take away his strike with your forward point, in this, strongly bind down on his halberd, and with this shoot up, simultaneously step inside twice toward him, and let your halberd fall, seize with your right hand down onto his left foot, and smash with your left [hand] forcefully onto his knee, then you may break his leg or throw [him].

If he has you held thus, then set your halberd onto his neck, and with this strongly press down, then you will break his work.

---

1. The Classical definition of *molam* is "jawbone," but the illustration plainly shows the left hand on the knee.

2. As with the previous plate, the illustration and physical mechanics of this technique suggest that *pedem* refers to the knee or leg, not the foot.

Cod. Vinob. 10826, 120r
The crossed guard against the lower.

    For this technique, position yourself in this manner: you should place the left foot forward toward the enemy, and thrust up from a low guard with a straight thrust to the adversary's eyes.

    But if he employs the same thing, while you are standing in the crossed guard, and the right foot leading against the enemy, then set the left to the inside. To be sure, during this step, you should strike against his attack from above; afterward, if you push down, you will displace the enemy's thrust.

    However, if he tries the same thing, them you should wind the enemy's halberd back with the blade of your halberd, from your left side toward the adversary's left side, and with this technique you will trap the enemy; at the same time, having seized the halberd, you should attack his eyes or chest.

    But if he attacks by stabbing you with a like technique, then you should displace the enemy's attack to the left with the blade of your halberd; afterward, you should strike the blade into his head, and withdraw from him with a double-step.

Cod. Vinob. 10826, 120r
The crossed guard against the lower.

    Note, hold yourself thus with this technique in the approach: step with your left leg toward him inside, and go from the lower guard up with a thrust into his face.

    If he thrusts thus at you from the lower, and you stand in the crossed guard against him, with your right foot forward, then step with your left leg inside, and in this step strike against his thrust from above, press his down with this properly, [and] then you will take his thrust away.

    If he strikes then at you thus, and he has your thrust engaged, then wind with your blade on his halberd from your left onto his left side, [so] then you will block him; in this, then twitch the halberd and thrust him to his face or the breast.

    If he thrusts thus at you, then take him thus aside with your blade onto your left side, simultaneously strike him with your blade to his head, [and] wind yourself with this twice from him to withdraw.

POLEAX

Cod. Vinob. 10826, 149r
A parrying form against a deadly strike using a poleax.

Place the right foot forward against the enemy during the approach, hold the poleax above your head with both hands, and the point extended toward the enemy. And if he places the left foot forward against you, and attempts to strike your head with a death blow, then you should displace him on the left side using your poleax; from that point, by drawing the poleax back, you should also pierce his groin.

However, when the enemy employs the same technique, you should parry this attack with the back point of the poleax. Afterward, following with the right foot, you should also stab his face with the blade of your poleax. However, if the adversary parries this thrust on the haft between both hands, you should wind the back point between both arms into the enemy's face.

If he attacks you in this same way, bring the right foot back, and parry the enemy on the haft between both hands. But if it should happen that he parries your attack, you should suddenly go for the upper openings by striking.

However, when the enemy tries to attack you by striking, you should parry him with the hilt of your poleax; from that point, if you have been pulled onto your right side, thrust into his face with the point; you should also withdraw while winding the poleax.

C94, 164r
A taking-off at the murder-ax against a murder-stroke.

Note, position yourself thus in this technique, when you travel together, stand with your right foot forward, and hold your murder-ax in both hands up high over your head, your point extended toward the opponent. If he stands then with his left foot against you, and strikes a murder-strike to your head, then take him aside with your murder-ax onto your left side, simultaneously pull your murder-ax to yourself, and seek toward his privates.

If he seeks below at you thus, then take him aside with your back point of your murder-ax, in this, pass forward with your right leg, and strike him with your blade of your murder-ax to his face. If he displaces you in-between both hands on his hilt, then wind your back point in-between the inside of both of his arms to his face.

If he strikes and winds you thus, then step with your right leg to the back and take him aside in-between both hands on the hilt. If he displaces you, then strike him quickly again to his upper opening.

If he strikes above at you thus, then take him aside with your hilt, pull with this onto your left side, and seek toward his face, in this, step and wind yourself with this from him to withdraw.

Cod. Vinob. 10826, 149v
A murderous strike against a thrust which punctures the chest.

It is necessary [that] you position yourself in the aforementioned fight in this way: you should stand with both legs straight, hold the haft of the poleax above your head with both hands, with the left raised behind your head, and at that moment, if you advance toward the enemy with the right foot, strike his head from above with the ax blade.

But if he employs a like technique against you, the left leading, and the right hand holding your poleax on behind the head, the left certainly brought toward the enemy to the point of the poleax from below, then parry the adversary's strike onto your right side with your ax's blade; at that moment, also thrust its front point into the enemy's chest.

However, if he tries to execute the same thing, you should displace his strike between both hands toward the left side; afterward, send the front point of your poleax straight above the left arm into the adversary's chest, and pull strongly while wrenching, and with this trouble taken care of, it will be possible to withdraw while winding the poleax.

C94, 164v
A murder-strike against a breast thrust.

Note, when you travel together, then position yourself thus in this technique, stand with equal feet upright, and hold your murder-ax with both hands on the hilt, your left hand up high behind your head, in this, step inside with your right leg, and strike him with your blade from above to his head.

If he strikes from above at you thus, and you stand with your left foot against him, and you hold the murder-ax with your right hand behind your head, your left by your forward point toward the opponent, then take away his strike with the blade [of] your murder-ax onto your right side, simultaneously thrust him with your forward point to his breast.

If he thrusts toward you thus, then take him aside in-between both hands on the hilt onto your left side, in this, wind him [with] your forward point [of] your murder-ax over his left arm to his breast, and pull strongly to you, step and wind yourself with this from him to withdraw.

Cod. Vinob. 10826, 150r
Two shoot-ins with the poleax.

In a mutual bind, you should place the left foot forward, hold the poleax with both hands, with the left on the haft near its blade, the right by the back point, and with a thrust having been shot over the enemy's left arm, you should turn the front point to his throat.

But if he attempts the same thing, you having placed the right foot forward, set his strike onto the left side with the blade of your poleax. Moreover, simultaneously following with the left, you will also bring the ax blade to his throat from the right side, and from this position, you should pull the adversary toward you.

However, when the enemy attempts the same thing against you, then you should displace this attack with the blade of your ax onto your right side, and thrust the forward point into the enemy's eyes; furthermore, you should immediately withdraw from him with the left.

However, if he stabs you with the same technique, following with the right, push his back point onto the left side with a thrust, then wind the poleax twice, strike the ax into his head, and thus you may withdraw from the enemy.

C94, 165r
Two break-ins by the murder-ax.

Note, hold yourself thus with this technique, when you travel together, stand with your left foot forward and hold your murder-ax in both hands, your left forward by your blade, your right by your back point, in this, break him with your forward point, and thrust in over his left arm to his throat.

If he thrusts at you thus, and you stand with your right foot against him, then take him aside with your blade onto your left side, in this, pass forward with your left leg and set your murder-ax's blade behind his neck [on] his right side, pull in with this to you.

If he is you thus to break-in and pulls you to him, then take him thus aside with your murder-ax's blade onto your right side, simultaneously thrust your forward point to his face and step with your left leg to the back.

If he thrusts toward you thus, then pass forward with your right leg and take him aside with your back point onto your left side, simultaneously double and strike him a murder-strike to his head, step and wind yourself with this from him to withdraw.

Cod. Vinob. 10826, 150v
A winding against a pulling technique.

In this fight, you should place the right foot forward, and hold the poleax against the enemy in your protected guard; then, if you follow with the left, you should strike a murderous blow to his head with the poleax.

However, if the adversary tries the same thing against you, holding your poleax with both hands, you will bring the right to the front of the haft, and the left to the back, and if you place the right foot forward, parry toward the left side with the haft, then follow to the inside with the left, thrust the back point over both arms, and during this action, having changed the left, pull the enemy's right hand from above with the blade of the ax, and his left toward you from below with the left arm.

But if he pulls you onto your left side in a like manner, you should remove the right from your poleax, and if you apply it to the enemy's left shoulder, you should push him away from you.

If he employs the same technique, bring the left foot back, and strike the forward point into his chest with a thrusting technique; moreover, then withdraw in your guard.

C94, 165v
A wind-in against a pull.

Note, when you travel together, then hold yourself thus with this technique, stand with your right foot forward, and hold your murder-ax in a good defense against the opponent, simultaneously pass forward with your left leg and strike him a murder-strike to this head.

If he strikes above at you thus, and you have your murder-ax in both hands, your right forward, your left by your back point, the right leg set in front, then take him aside with your hilt onto your left side, in this, pass forward with your left leg, and wind him your back point over both of his arms, with this exchange your left hand to your back point and pull up his right hand with the blade and his left hand down with your left arm to you.

If he pulls you thus to him onto his right side, then allow your right hand loose from your murder-ax, set him with this onto his left shoulder and push him from you.

If he pushes you from him thus, then step with your left leg to the back, and wind your forward point with a thrust to his breast, step with this from him to withdraw into a good defense.

Cod. Vinob. 10826, 151r
A thrust which attacks the groin against a wind-in.

It is necessary that you position yourself thusly in this fight: you will place the left foot forward, [and] you should hold the poleax raised in front of your face, with the right placed in the middle of the haft, the left by the back point; in that instant, if you follow with the right foot, you should draw up into the enemy's groin with the front point.

But if he employs the same thing against you, the right leading, and holding the poleax in front of your face with both hands, with the right placed at the back point above and the left near the front, then you should displace his attach onto the left side with the back point of your poleax, and in that instant cut the blade into his head by striking. And if he parries your attack between both hands with the haft, following with the left foot, you should wind the back point between both of the adversary's arms into his face. And if he parries again, having brought the left foot back, you will strike his arm with your ax's blade; at that moment, you will remember to withdraw while winding the poleax.

C94, 166r
A groin thrust against a wind-in.

Note, position yourself thus into this technique, when you travel together, stand with your left foot forward, and hold your murder-ax in the high [guard] by your face, your right in the middle of your hilt, your left by your back point, in this, step inside with your right leg, and thrust your forward point to his groin.

If he thrusts toward you thus, and you stand with your right foot against him, and you hold your murder-ax in both hands by your face, your right below by your back point, your left above by your forward, then take him aside with your back point onto your left side, simultaneously strike him with your blade to his head. If he displaces you in-between both hands on his hilt, then pass forward with your left leg and wind him your back point with a thrust in-between the inside of both of his arms to his face. If he goes to displace you thus, then step with your left leg to the back, and strike him with your blade to his right arm, wind yourself with this from him to withdraw.

Cod. Vinob. 10826, 151v
Two methods of pulling, from which a throw is made.

If you henceforth want to win the prize, you will position yourself in this way: place the right foot forward, hold the poleax raised in front of your face on the left side—the right should hold the back point, and the left the middle of the haft—then, follow with the left foot, and cut the enemy's head with the blade of the ax. If the enemy parries you between both hands on the haft, then thrust the front point with the blade attached above the right knee, and pull toward you.

But if he in turn employs the same thing, then you in turn will not hesitate to apply the blade to the enemy's own neck from the position of a strike, on the right side of the neck, and if you pull the enemy strongly from this position, that man's attack will not be able to harm you.

And if the adversary pulls you from above with a like technique, you should pass in a triangle, and if you parry his attack between both hands on the haft of the murder-ax, and also press up hard, you will render him weak from above. You should then be sure to remove the left hand from the poleax; if you bring it to his right foot, and lift up, you will throw the enemy onto his back.

C94, 166v
Two pulls from which goes a throw.

Note, when you travel together, then position yourself thus into this technique, stand with your right foot forward and hold your murder-ax in the high [guard] by your face on your left side, your right hand forward by your back point, your left on the middle of your hilt, in this, pass forward with your left leg and strike him with your blade to his head. If he takes you aside in-between both hands on his hilt, then wind him your forward point and blade below his right knee-bend, with this pull toward yourself.

If he has established you below thus, and pulls you to him, then go toward him also with your blade of your murder-ax with a strike around his throat [on] his right side, pull with this strongly to you, then he may not injure you with his position on the leg.

If he pulls you over to him thus, then step in a triangle and take him aside in-between both hands on your hilt, press with this strongly up, then when you take him up thus to fall back, in this, loose your left hand from your murder-ax, seize him with this [on] his right leg, and lift it up, then you will throw [him] onto his back.

Cod. Vinob. 10826, 152r[1]
A method of entry using the poleax.

If you approach near to the enemy, you should stand with both legs straight, [and] you should hold the poleax raised in front of your face; then, if you pass inside with the left, strike the enemy's head with the blade.

But if the adversary employs the same thing against you, then you should strike against his blow, and parry the enemy's strike in this way. Moreover, you should at that moment thrust the lower point over his right arm, and you should apply your right to the adversary's poleax; then, if you turn yourself onto your right side, you will press the enemy's right between both poleaxes, not without his suffering.

But if he uses this same technique, and wrenches the ax from you, you should give it up to the enemy, definitely grab his right armpit on the outside with the left hand, and from this position, strongly push the adversary away from you, but also snatch the poleax again from his hands by wrenching, and if you pass backward, you should strike a murder-stroke to the enemy's head with the blade.

C94, 17r
A break-in at the murder-ax.

Note, hold yourself thus with this technique, when you travel together, stand with equal feet up right, and hold your murder-ax in both hands in the high [guard] before your face, in this, pass forward with your left leg and strike him with your blade to his head.

If he strikes above at you thus, then strike to counter his strike taking him away with this, in this, break in with your lower point over his right arm, and catch his murder-ax with your right hand together with yours, turn yourself with this onto your right side, then you will press on his right hand in-between both murder-axes.

If he is to break you thus, and wants to take your murder-ax, then you will let him, in this, seize him with your left outside under his right shoulder, press in with this strongly from you, and with your right seize again on your murder-ax, simultaneously pull yours out of his hand, step with this to the back and strike him a murder-strike with your blade to his head.

---

1. This plate is misnumbered "151."

Cod. Vinob. 10826, 152v
A murder-stroke against a method of pulling.

You should position yourself in this type of fight thusly: you will place the right foot forward, hold the haft of the poleax with both hands, and strike a murder-stroke to the adversary's head with the blade of your ax.

But if he attempts this same thing against you, left leading, then strike the front point between the enemy's arms, and if you hook his neck from the right side, you should parry his strike with the back point, then pull the enemy toward you with the curved blade.

And if he employs a like technique against you, then you will move the right from the inside onto the outside of the poleax, and from this position you should drive the enemy's ax downward by pressing down, and you should also cut the left side of the enemy's neck with the blade.

But if the adversary uses the same thing, you should displace that attack with the poleax toward your right side; moreover, if you immediately thrust the front point into his eyes, you should withdraw from the enemy.

C94, 167v
A murder-strike against a pull.

Note, when you travel together, then hold yourself thus in this technique, stand with your right foot forward, and have your murder-ax with both hands on your hilt, in this, strike him a murder-strike with your blade to his head.

If he strikes above at you thus, and you stand with your left foot against him, then go to him with your forward point in-between the inside of both of his arms, setting it with this onto the right side of his throat, and take away his strike with your back point, simultaneously pull him with your hook [of] your blade toward you.

If he pulls you to him thus, then from inside quickly exchange your right hand outside quickly around on your murder-ax, with this wind his murder-ax down, and cut him with your blade to the left side of his throat.

If he has you established thus, then take him aside with your murder-ax onto your right side, in this, thrust him with your forward point to his face, step and wind yourself from him with this to withdraw.

Cod. Vinob. 10826, 153r[1]
An over-strike against a thrust.

In this mutual bind, place the left foot forward, and hold the poleax raised in front of your face with both hands, then following with the right, you will strike the enemy's head with your ax's blade.

But if he works the same thing against you, the left leading, holding the front point of the poleax against the enemy with the left hand, and the right at the back, displace the enemy's strike with the haft between both hands; at that moment, stab his throat with the front point.

However, if he attacks you with the same technique, then if you set this attack onto your right side with the front point, thrust straight over the left arm into the enemy's face. But if he parries this attempt with the poleax, then thrust the back point into his left side.

When he employs this same thing, parry this attack with the blade, afterward following with the left, drive the front point between both his hands into the enemy's chest, and you should withdraw from him while winding the poleax.

C94, 168r
An upper-strike against a thrust.

Note, position yourself thus into this technique, when you travel together, stand with your left foot forward and hold your murder-ax in both hands in the high [guard] before your face, in this, pass forward with your right leg and strike him with your blade to his head.

If he strikes above at you thus, and you stand with your left foot against him, and hold your murder-ax with your left hand by your forward point toward the opponent, your right by your back, then take away his strike in-between both hands on your hilt, in this, thrust him with your forward point to his throat.

If he thrusts toward you thus, then set him aside onto your right side, in this, thrust him with your forward point over his left arm to his face. If he displaces you with his murder-ax, then wind your back point to his left side.

If he winds below thus, then set him aside with your blade, in this, pass forward with your right leg and thrust your forward point inbetween both of his arms to his breast, step and wind yourself with this from him to withdraw.

---

1. This plate is misnumbered "152."

Cod. Vinob. 10826, 153v
A position or guard, whereby you hold the poleax applied to the chest.

You should position yourself in this fight with this technique: you should place the left foot forward, and hold the poleax with both hands, the left applied to the the front part of the staff, toward the enemy, the right in fact brought to the back point and joined to the chest; at that moment, follow with the right foot, and from the position of a strike which is named after the wind, you should steer the poleax from your chest toward the enemy's head.

However, if he approaches you with the same technique, the right leading, the back part of the poleax held toward the enemy with the right, and the front with the left on your right side, then with the right foot brought back, you should parry the enemy's attack between both hands on the haft, meanwhile also position the right foot forward again, and steer the back point into his face.

If the adversary parries it on the poleax between two hands, following with the left, strike the blade fatally into his head, and you will take care to withdraw from the enemy while protected by rotating the poleax.

C94, 168v
A breast-guard against a take-off.

Note, when you both at the same time travel, then position yourself thus in this technique: stand with your left foot forward, and hold your murder-ax in both hands, your left by your forward point toward the opponent, your right below by your back point over your breast; in this, pass forward with your right leg and wind him with an wind-strike from your breast to his head.

If he strikes over to you thus, and you stand with your right foot toward him, and hold your murder-ax with your right hand by your back point, toward the opponent, your left by your forward on your left side, then step with your right leg to the back, and displace his strike in-between both hands on your hilt; in this, step with your right leg again inside, and wind him your back point to his face.

If he displaces thus in-between both hands on his murder-ax, then pass forward with your left leg and strike him a murder-strike with your blade to his head, step and wind yourself with this from him to withdraw.

Cod. Vinob. 10826, 154r[1]
Two windings of the back point.

You should in turn place the right foot forward in this clash, [and] hold the poleax with both hands, the right applied to the back point, toward the enemy, and the left near the blade, then from this position, if you follow with the left foot, you should strike the enemy's throat with the back point having been thrust between his arms.

But if the adversary employs the same thing against you, standing on equal legs, and also holding the poleax in the middle with both hands raised in front of your face, then you should set the enemy's attack toward your right side with the back point. Afterward, follow with the left, and you should strike the back point between his arms toward the adversary's face. But if he parries this attempt, you should suddenly strike the blade into his head with all your might.

And if the enemy attacks your head with a like technique, you should bring the left back, and parry his attack on the haft between both hands, at that moment press the adversary's poleax downward with the blade of your poleax, and if you thrust the front point into his eyes, you should withdraw from the enemy.

C94, 169r
Two wind-ins with both back points.

Note, hold yourself thus with this technique, when you both at the same time travel, stand with your right foot forward, and hold your murder-ax in both hands, your right by your back point toward the enemy, your left back by your blade; in this, pass forward with your left leg, and wind him your back point in-between the inside of both arms to his throat.

If he winds you thus, and you stand with equal feet on your right against him, and you hold your murder-ax with both hands in the middle of your hilt in the high [guard] before your face, then take him thus aside with your back point onto your right side; in this, pass forward with your left leg and wind him your back point inbetween the inside of both arms to his face. If he displaces you thus, then strike him quickly a murder-stroke with your blade to his head.

If he strikes over to you thus, then step with your left leg to the back and displace him thus in between both hands on your hilt; in this, pull him with your blade, his murder-ax under it, and thrust him with your forward point to his face, and wind yourself with this from him to withdraw.

---

1. This plate is misnumbered "153."

Cod. Vinob. 10826, 154v
A thrust from which the chest is attacked, against a middle strike.

If you approach [each other] mutually, you should position yourself in this technique: you will not hesitate to place the left foot forward, you should hold the poleax with the left near the blade toward the adversary, with the right also applied to the back point at your head, then following with the right, you should strike the front point to the chest.

And if the adversary attacks you from above with a like technique, the right leading, holding the poleax on the left side with two hands, you should lift the back point up, and parry the enemy's thrust on the left side, then also following with the left foot, you should cut the adversary's right side with the blade from the position of a middle strike.

However, if the enemy does the same thing, you should parry his attack with the haft of your poleax; meanwhile, you should push the enemy's poleax down by its blade onto your left side, and jab his chest. And if he parries it between both hands on the haft, you should immediately strike the enemy's head with the blade, and then withdraw from him.

C94, 169v
A breast-thrust against a middle-strike.

Note, when you both at the same time travel, then hold yourself thus with this technique, stand with your left foot forward and hold your murder-ax with your left hand by your blade toward the opponent, your right by your back point behind your head, in this, pass forward with your right leg and thrust him with your forward point to his breast.

If he thrusts thus from above at you, and you stand with your right foot foot toward him and hold your murder-ax in both hands on your left side, then go with your back point on and take his thrust away onto your left side; in this, pass forward with your left leg and strike him a middle-strike to his right side with your blade.

If he strikes you thus to your right side, then take him thus aside with your hilt [of] your murder-ax, simultaneously pull with your blade his murder-ax onto your left side under-his and thrust him to your breast. If he displaces you thus in-between both hands on the hilt, then strike him quickly with your blade to his head and step with this from him to withdraw.

Cod. Vinob. 10826, 155r[1]
A way of setting-aside against that strike which cuts the foot.

If the enemy comes into range, you should stand straight on both feet, and hold the poleax raised above the head with both hands; then, you should advance with the left from this position of the body, and strike with the blade from a high strike [guard] straight to his leading leg.

However, if you are leading with the left, and he also strikes that foot, you should lower the poleax, and place it in front of the left foot, the front point tilted toward the ground, and from this position, you should parry his strike. Then, the poleax having been raised, following with the right foot, you should thrust the front point into his neck. But if he sets this same attack onto the right side with the blade, you should send the point underneath, and stab the left flank.

However, if he attempts the same thing, you should parry onto the left side with the blade of your poleax; then, following with the right, you should guide both points between both arms into the enemy's face or chest; afterward, if you also strike his head with the blade, you should withdraw from him.

C94, 170r
A take-off against a strike to the leg.

Note, position yourself thus into this technique: when you both at the same time travel, stand with equal feet on the right, and hold your murder-ax in both hands in a high position over your head; in this, step with your left leg inside and strike him low from above with your blade to his forward leg.

If you stand then with your left foot toward him and he strikes after you, then for with your murder-ax below his, and set you fore your left leg your forward point onto the earth, then take him with this his strike aside; in this, go on with your murder-ax and pass forward with your right leg, [and] thrust him with this, your forward point to his throat. If he takes you thus aside with his blade on his right side, then change simultaneously quickly below through and thrust him to his left side.

If he thrusts thus toward your, then take him thus aside with your blade onto your left side; in this, pass forward with your right leg and wind him your back and forward point in-between the inside of both arms to his face or the breast, simultaneously strike him with your blade to his head, and step with this from him to withdraw.

---

1. This plate is misnumbered "154."

Cod. Vinob. 10826, 155v
A method of crossed parrying against a thrust which attacks the groin.

You will position yourself in this way: you should place the right foot forward, hold the poleax raised above your head with both hands, the right applied to the front point, toward the enemy, the left also at the back point; then, following with the left, you should draw up into the adversary's groin with the front point.

However, if he tries the same thing against you, the left leading, and you hold the poleax toward the enemy with crossed arms, with the left hand positioned near the blade of the ax, the right placed forward on the haft, [then] with the left foot brought back, you should set his thrust onto your left side with the back point; meanwhile, immediately advance again with the left, and you should guide the front point from underneath up into the enemy's face or chest.

When the adversary employs a like technique, parry it between both of your hands on the haft; then, if you pass in a triangle, cut his head with the blade, and with this technique having been used, withdraw from the enemy.

C94, 170v
A crossed taking-off against a groin thrust.

Note, when you both at the same time travel, then position yourself thus into this technique: stand with your right foot forward and hold your murder-ax in both hands over your head, your right by your forward point toward the opponent, your left by your back; in this, pass forward with your left leg, and thrust him with your forward point to his groin.

If he thrusts you thus to your groin, and you stand with your left foot toward him, and hold your murder-ax with crossed arms against the enemy, your left hand by your blade, the right forward on the hilt, then step with your left leg to the back, and take his thrust with your back point away onto your left side; simultaneously step quickly with your left leg inside again, and wind him your forward point from below on to his face or the breast.

If he winds you thus, then take him thus aside in-between both hands on your hilt; in this, step in a triangle, and strike him with your blade to his head; withdraw yourself with this from him to the back.

Cod. Vinob. 10826, 156r[1]
A deadly strike against a crossed press, so that he is open underneath.

It is necessary that you position yourself against the enemy in a mutual approach as follows: you will not hesitate to place the left foot forward, and hold the poleax against the enemy with the arms in the shape of a cross; then, from the position of a thrust, you will apply the front point of the poleax to the enemy's left arm.

However, if you are positioned against him [with] the left leading, and he attempts to stab the left arm, [then] with the left brought back, you will set the enemy's thrust onto your left side with the blade, and in fact, quickly stepping forward again, you should strike a death-stroke into his head with the blade.

But if he attacks you with a like technique, parry his attack between both hands on the haft, then also following with the right foot, extend the front point toward the adversary's face or chest. If he parries it, having brought the right foot back, if you attack his leading left foot with the strike which is named after the wind, you should withdraw from the enemy.

C94, 171r
A murder-strike against a crossed placement.

Note, hold yourself thus with this technique: when you both at the same time travel, stand with your left foot forward, and hold your murder-ax with crossed arms toward the opponent; in this, set him with your forward point onto his left arm with a thrust.

If you stand then also thus with your left foot toward him and he thrusts you to your left arm, then step with your left leg to the back and take him thus aside with your blade onto your left side; simultaneously step quickly again inside and strike him a murder-stroke with your blade to his head.

If he strikes over to you thus, then take him thus aside in-between both hands on your hilt; in this, pass forward with your right leg and wind him your forward point with a thrust to his face or the breast. If he displaces you thus, then step with your right leg again to the back and strike him a wind-strike to his left forward leg; step with this from him to withdraw.

---

1. This plate is misnumbered "155."

Cod. Vinob. 10826, 156v
Two presses of the poleax.

You will position yourself like so in this fight: place the right foot forward, and place both hands at the back point of the poleax, raised above the head; then, following with the left, strike the side of the enemy's head with a transverse stroke.

But if he attempts the same thing against you, the left forward toward him in the position of the scales, and you hold the poleax in both hands in such a way that the front point is leaning on the ground, [then] with the poleax raised, you should parry the enemy's transverse strike between both of your hands on the haft of your poleax.

But if you are parried from the enemy in the same way, then apply the front point of the poleax to his left arm.

However, if he employs the same thing against you, you should parry it with the blade, and at the same time trap the enemy's poleax with the blade of your ax, and if you drive the front point into his body or cuirass with all your might, you will force the adversary to fall with this thrust.

C94, 171v
Two placements at the murder-ax.

Note, when you both at the same time travel, then hold yourself thus with this technique: stand with your right foot forward, and hold your murder-ax in both hands by your back point in the high-position over your head; in this, pass forward with your left leg and strike him a thwart-strike to his right side [of] his head.

If he strikes over to you thus and you stand with your left foot toward him, properly in the scales, your murder-ax in both hands, your forward point toward the earth, then go on with your murder-ax and take his thwart-strike between both hands on your hilt away.

If he has you also thus removed, then set him with your forward point [of] your murder-ax with a thrust onto his left arm.

If he has you thus established, then take him thus aside with your blade, simultaneously block his murder-ax with your blade and thrust him with your forward point; afterward, press with strength into his body, then you will thrust him to the earth.

VARIOUS WEAPONS

Cod. Vinob. 10826, 15r
A form of fighting with the short spear, which foreigners use against the longsword.

For the technique of the short spear, you will position yourself in this manner: you should place the left foot forward; you should hold it on the left side with both hands, the front point extended toward the enemy. And if he also stands against you with the left leading, and raises the sword in the guard of a high strike, and attempts to strike the head, then quickly thrust the point of the short spear into the enemy's body.

However, if he attempts the same thing against you, bring the left foot back, and you should strike his spear onto the adversary's left side with a cross-[strike], so that with this technique you will repel his attack.

But when this same thing is employed against you, impeding your attempt, following with the right foot, if you drive into the right side of the enemy's body with the front point of the spear, you will be able to withdraw from him.

C93, 227r
The short spear against the sword.

Note, position yourself in this technique with your short-spear: stand with your left foot forward and hold it in both hands on your right side, your forward point, or spear, toward the opponent. If he stands then also with his left foot toward you and holds his sword in an upper-strike toward you in the high guard and wants to strike you to your head, then thrust him quickly with your forward point of your short spear to his body.

If he thrusts at you thus, then step with your left leg to the back and strike him with your sword to his staff [of] his short spear on his left side with the crooked [strike], then you will take his thrust with this away.

If he strikes you thus together and takes your thrust away, then pass forward with your right leg and thrust him with your forward point [of] your short spear to his body's right side; step with this from him to withdraw.

Cod. Vinob. 10826, 15v
The halberd against the longsword

For this technique of wielding the halberd, if you wish to compose yourself properly and in the way of athletes against an enemy using a sword, you will remember to place the left foot forward, [and] you should hold the halberd with the right hand near the back point, toward the enemy, the left applied to the middle of the halberd directly up in front of your face; from your left side, you should then quickly strike from this position into the adversary's head.

But if the enemy attacks you from above in the same way, [your] left foot also leading, and [you are] holding the sword above the right shoulder, the knot of the sword in the left hand, toward the enemy, if you strike his halberd with the long edge sent out, his strike will be found ineffective. Moreover, if you immediately pass in a triangle from this position, you will strike his head by sending the long edge out from the left side.

But if he tries the same thing against you, [then] following by pressing forward with a double step, you will attack by thrusting toward him twice, from above and below, and then you should withdraw from the enemy.

C93, 227v
The halberd against the sword.

Note, when you both at the same time travel, then position yourself thus in this technique with your halberd: stand with your left foot forward and hold your halberd with your right hand by your back point, toward the opponent, your left in the middle extended before your face in the high position on your left side; simultaneously strike him quickly to his head.

If he strikes over at you thus, and you also stand with your left foot toward him and hold your sword on your right shoulder, the pommel in the left hand toward the opponent, then strike him with the long edge to his halberd; with this, you will then take his strike away; simultaneously step in a triangle and strike him with the long edge to the left side of his head.

If he strikes over at you thus, then step twice inside with an after-attack and thrust at him below and above twice; step with this from him to withdraw.

Cod. Vinob. 10826, 13r
The lance against the sword.

You will prepare yourself toward the enemy in the fight as follows: you will place the left foot forward, you should hold the lance toward the enemy with the left hand forward, you will place the right hand behind the right foot on the right side, and from this position of the body, attack his eyes.

But if he uses the same thing against you, [your] right foot leading, and your sword is in the guard of the change on the left side, the point tilted toward the ground, [then] if you raise it, and you strike his lance, the adversary's attack will be repelled in this way onto your right side.

But if he displaces your lance with a strike of his sword in this way, guide the lance down, but then you will immediately lift it up, strongly grip the lance in the middle, and thrust through the left flank of the adversary's body.

However, if he attacks you with the same technique, then violently strike the long edge of your sword into his head with all your might, and if this strike does not happen as quickly as possible, you will suffer in the back parts.

C93, 228r
The long spear against the sword.

Note, hold yourself thus with this technique: when you both travel at the same time, stand with your left foot forward and hold your spear with your left hand forward, toward the opponent, your right on your spear behind your right leg on your right side, [and] simultaneously thrust him quickly to his face.

If he thrusts over at you thus, and you stand with your right foot toward him and hold your sword in the changer on your left side, the point toward the earth, then go on with your sword from the changer, strike him with this to his spear, [and] then you will take his thrust away onto your right side.

If he has struck toward your spear thus, then withdraw your spear [from] under his; in this, go again from this holding your spear well in the middle before him, and thrust him through the left side of his body.

If he has you thus through to thrust through your body, then strike him with strength, with your long edge to his head.

Cod. Vinob. 10826, 13v
A technique of the halberd against an enemy who fights you with a sword.

If you want to athletically employ the aforementioned technique, you will place the right foot forward, you should hold the halberd set on the right arm, with the right hand near its weaker part, the left also in the middle against the enemy, [and] the point should tilt up; then, thrust the front point into the eyes.

However, if the adversary attacks you with the same technique, [you] standing in the balance, and you also place the left foot forward, you should hold the hilt of the sword with both hands on your left side, the point extended toward the enemy; press the long edge of your sword to his halberd, and from this position you will repel the enemy's attack. However, if you execute it, step inward with the right, and with the sword raised above your head with your hands formed in the shape of a cross, you should strike by sending the long edge into the enemy's head [as depicted by the right figure].

However, when the adversary attempts the same thing, you should extend the halberd forward, [it] having been positioned back behind the head, and in turn strike his head with the halberd's blade [as depicted by the left figure].

C93, 228v
The halberd against the sword.

Note, when you both at the same time travel, then hold yourself thus with this technique: stand with your right foot forward and hold your halberd with your right hand on your lower point on your right hip, your left in the middle toward the opponent, the point above his, [and] simultaneously thrust him with your forward point to his face.

If he thrusts over at you thus and you also stand with your left foot toward him, properly in the "scales," and hold the sword with both hands on your hilt on your left side, the point toward the opponent, then set him with your long edge [to] his halberd and set him thus with this aside; simultaneously step with your right leg inside and go on with [crossed] arms with your sword over your head, and strike him with the long edge over to his head.

If he strikes thus over at you, then wind your halberd behind your head to drive and strike him also with your blade of your halberd to his head.

Cod. Vinob. 10826, 19r
A technique of the boar-spear against the halberd.

If you both meet in a fight, and you then wish to take a prize out of the gymnasium, you will remember to place the left forward; you should also hold the boar-spear on the right side with the right hand, with the left placed in the middle of the boar-spear, toward the enemy, and from this position of the body, you will thrust into the enemy's left flank.

But if he tries the same thing against you, [your] right leading, and holding the halberd on the right side with the right hand near its back part, the blade toward the enemy, then you should repel the enemy's attack onto your right side with the blade of your halberd, and immediately following with the left, strike the adversary's groin with a thrust.

However, if the enemy attempts the same thing, you should bring the left back; you will parry the enemy's attack onto the left side with the front part of the boar-spear.

But if he attempts to parry you with the same technique, strike your halberd's blade into his head.

But should it happen that he attacks you from above in the same way, you will meet the enemy's strike, and if you thrust through the enemy's throat by guiding the spearhead over his left arm, the unskilled adversary will be forced to lose his life.

C93, 232r
The boar-spear against the halberd.

Note, hold yourself thus with this technique: when you both travel at the same time, stand with your left foot forward, and hold your spear with your right hand on your right side, your left in the middle of your staff toward the opponent; simultaneously thrust toward his left side of his body.

If he thrusts at you thus, and you stand with your right foot toward him, and hold your halberd with your right hand by your back point on your right side, your blade toward the opponent, then take him thus aside with your blade on your left side, simultaneously pass forward with your left leg, and thrust him to his groin.

If he thrusts at you thus below, then step with your left leg to the back, and take him thus aside with your forward waist of your spear on your left side.

If he has you thus engaged, then strike him with your blade [of] your halberd to his head.

If he strikes over at you thus, then go on toward his strike and thrust him over his left arm through his throat; then, you will thrust him to the earth.

Cod. Vinob. 10826, 19v
Another form of combat with the boar spear against the halberd.

When you want to use this aforementioned technique against an advancing enemy, you should place the left foot forward, you will not hesitate to hold the boar-spear with the right near the lower point, the left applied to the middle, in the air in front of your vision, and you will strongly strike the adversary's head.

But if he in turn tries to strike your head, the left also set in front of you, and with the right gripping the back point of your halberd behind the head, the left also applied to the same thing, near the blade, against the enemy, then you want to displace his attack with the blade of the halberd toward your right side. To be sure, quickly pierce the enemy's chest with the front point.

And if he attempts to stick you, repel the adversary's thrust onto the right side with an attack of the boar-spear; meanwhile, also step to the inside with the right foot, with the boar-spear having been cast away, control the right shoulder by guiding the left hand around the enemy's neck, the right certainly [around] his left foot, and if you lift the enemy up from this position, you will throw him prone, so you will treat the thrown man according to your will.

C93, 232v
More of the boar spear against the halberd.

Note, when you both travel at the same time, then hold yourself thus with this technique: stand with your left foot forward and hold your spear with your right hand on your lower point, your left in the middle in the high position before your face; simultaneously strike him to his head.

If he strikes over at you thus, and you also stand with your left foot toward him, and hold your halberd with your right hand by your back point behind your head, your left on your staff, by the blade, toward the opponent, then take him thus aside with your blade onto your right side; simultaneously thrust him with your forward point to his breast.

If he thrusts at you thus, then take him thus aside with your spear onto your right side; simultaneously step with your right leg inside; allow your spear to fall with this; in this, seize him with your left hand behind his throat around onto his right shoulder, and with your right to his left leg, lift him up with this strongly, then you will throw him onto the face. If you have him thus thrown, then you will want to grasp your spear again and deal with him as you wish with strikes or thrusts.

Cod. Vinob. 10826, 120v
A technique of the halberd against the dussack

Position yourself for this technique as follows: step toward the enemy with your left foot, and cut his head with a high strike.

And if the adversary attacks you from above in a like manner, [your] right leading, then following with the left, you should raise the dussack in the position of the crown, and parry the enemy's strike.

However, if he parries your attack with this same technique, then advance toward him with the right foot, and you should also wind the lower point from below into the enemy's eyes or chest.

But if he winds the point toward you in a similar way, then from the position of the crown, you will strike the dussack over his halberd, and push down, and in this manner the enemy's force is displaced. Meanwhile, you should also seize the back point of the adversary's halberd with the left hand, and accordingly, grip the halberd firmly; then, following with the right, quickly strike his head.

But if he parries this attack, then stab his chest or face, and you should withdraw after a double-strike.

Cod. Vinob. 10826, 120v
A technique of the halberd against the dussack.

Note, position yourself thus in this technique with the approach: step with your left leg inside and strike him a upper-strike to his head.

If he strikes at you thus, and you stand with your right foot forward, then step with your left leg inside and strike with the dussack properly up into the crown before your head, displacing his strike with this.

If he has your strike thus engaged, then step with your right leg inside, and wind him [with] your forward point from below into his face or breast.

If he has you thus with his point [wound], then fall to him from the crown onto his halberd, and press with this properly again; then, you will take him thus aside, in this, seize him with your left hand on his back point, hold with this his halberd, and step with your right leg inside, and strike him simultaneously to his head.

If he then displaces you thus, then thrust him to his face or the breast, simultaneously withdraw yourself with a double-strike from him to the back.

Cod. Vinob. 10826, 121r
Another of the aforementioned techniques.

In this fight, you should place the left foot forward, and pierce the enemy's chest with the halberd.

But if he attempts to stick you with in a like manner, [your] left leading, then you should lift your dussack in the position of the Crown, and you should parry the enemy's thrust; meanwhile, you should sieze the middle of his halberd with the left hand, and in this way you should push to the left, and also thrust the point of your dusack into the enemy's eyes.

However, if he attacks you with a like technique, then you should parry to the left between both hands on the halberd.

But if he parries the same thing, you will strike the adversary's head by sending your long edge forward, and you should withdraw from him by thrusting and striking with a double-step.

Cod. Vinob. 10826, 121r
One other technique of the halberd against the dusack.

Note, hold yourself thus with the approach in this technique: step with your left leg inside, and thrust him [with] your halberd to his body.

If he thrusts thus at you, and you stand with your left foot toward him, then strike up with your dussack, into the crown, take his thrust away with this, simultaneously seize with your left hand in the middle of his halberd, press his down with this well onto your left side, toward the earth, [and] simultaneously thrust him with your point [of] your dussack to his face.

If he thrusts thus at you, then take him thus aside in-between both hands on your halberd, onto your left side.

If he has you thus engaged, then strike him with your dussack with the long edge to his head, and wind yourself with thrusts and strikes twice from him.

Cod. Vinob. 10826, 121v
A technique with the Spanish sword against the boar-spear.

During a mutual approach, you will position yourself in this way: you will place the left foot toward the enemy, and thrust the boar-spear to the enemy's eyes or chest.

But if he employs this same technique, holding the Spanish sword with the right foot placed in front of you, then you should strike against his thrust, and bring [the weapon] into contact near the tip of the boar-spear. Meanwhile, if you press to the left, you have averted his thrust, and afterward you should sieze the boar-spear on the forward part.

But if he attempts to sieze yours, with the boar-spear pulled back, from the enemy's left side to the right, thrust to the pectoral or eyes.

But if the adversary attacks you with a like technique—that is, with a double thrust—then you should displace to the right with the Spanish sword; afterward, you should pass to the inside with the left foot, and sieze the middle of his spear, and quickly stab his eyes. However, if he parries, then you will change through from his spear; at the same time, if you leap forward in a triangle, you should strike the enemy's head.

Cod. Vinob. 10826, 121v
A technique of the rapier against the boar spear.

Note, position yourself thus in this technique with the approach, step with your left leg to him inside, and thrust him with your boar spear to his face or the breast.

If he thrusts thus at you, and you stand with your right foot toward him with your rapier, then strike against his thrust setting it with this forward onto his spear, simultaneously press from you onto your left side, then you will take his thrust away, seize him with this forward to his spear.

If he wants your spear thus to seize, then withdraw [it] quickly to you, and thrust him from his left to his right side, to his breast or the face.

If he thrusts thus twice at you, then take him thus aside onto your right side with your rapier, in this step with your left leg inside, and seize with your left hand in the middle of his spear, and thrust him simultaneously quickly to his face. If he displaces you thus, then change through on his spear, and spring in a triangle and strike him to his head.

Appendix D

Register

Vienna	Dresden	Plate	Page
Cod. Vinob. 10825, 154r	C93, 183r	Shortstaff 1	8
Cod. Vinob. 10825, 154v	C93, 183v	Shortstaff 2	10
Cod. Vinob. 10825, 155r	C93, 184r	Shortstaff 3	12
Cod. Vinob. 10825, 155v	C93, 184v	Shortstaff 4	14
Cod. Vinob. 10825, 156r	C93, 185r	Shortstaff 5	16
Cod. Vinob. 10825, 156v	C93, 185v	Shortstaff 6	18
Cod. Vinob. 10825, 157r	C93, 186r	Shortstaff 7	20
Cod. Vinob. 10825, 157v	C93, 186v	Shortstaff 8	22
Cod. Vinob. 10825, 158r	C93, 187r	Shortstaff 9	24
Cod. Vinob. 10825, 158v	C93, 187v	Shortstaff 10	26
Cod. Vinob. 10825, 159r	C93, 188r	Shortstaff 11	28
Cod. Vinob. 10825, 159v	C93, 188v	Shortstaff 12	30
Cod. Vinob. 10825, 160r	C93, 189r	Shortstaff 13	32
Cod. Vinob. 10825, 160v	C93, 189v	Shortstaff 14	34
Cod. Vinob. 10825, 161r	C93, 190r	Shortstaff 15	36
Cod. Vinob. 10825, 161v	C93, 190v	Shortstaff 16	38
Cod. Vinob. 10825, 162r	C93, 191r	Shortstaff 17	40
Cod. Vinob. 10825, 162v	C93, 191v	Shortstaff 18	42
Cod. Vinob. 10825, 163r	–	Shortstaff 19*	44
Cod. Vinob. 10825, 163v	–	Shortstaff 20*	46
Cod. Vinob. 10825, 164r	–	Shortstaff 21*	48
Cod. Vinob. 10825, 164v	–	Shortstaff 22*	50

Vienna	Dresden	Plate	Page
Cod. Vinob. 10825, 166r	C93, 194r	Longstaff 1	54
Cod. Vinob. 10825, 166v	C93, 194v	Longstaff 2	56

Cod. Vinob. 10825, 167r	C93, 195r	Longstaff 3	58
Cod. Vinob. 10825, 167v	C93, 195v	Longstaff 4	60
Cod. Vinob. 10825, 168r	C93, 196r	Longstaff 5	62
Cod. Vinob. 10825, 168v	C93, 196v	Longstaff 6	64
Cod. Vinob. 10825, 169r	C93, 197r	Longstaff 7	66
Cod. Vinob. 10825, 169v	C93, 197v	Longstaff 8	68
Cod. Vinob. 10825, 170r	C93, 198r	Longstaff 9	70
Cod. Vinob. 10825, 170v	C93, 198v	Longstaff 10	72
Cod. Vinob. 10825, 171r	C93, 199r	Longstaff 11	74
Cod. Vinob. 10825, 171v	C93, 199v	Longstaff 12	76

Vienna	**Dresden**	**Plate**	**Page**
Cod. Vinob. 10825, 178r	C93, 202r	Halberd 1	80
Cod. Vinob. 10825, 178v	C93, 202v	Halberd 2	82
Cod. Vinob. 10825, 179r	C93, 203r	Halberd 3	84
Cod. Vinob. 10825, 179v	C93, 203v	Halberd 4	86
Cod. Vinob. 10825, 180r	C93, 204r	Halberd 5	88
Cod. Vinob. 10825, 180v	C93, 204v	Halberd 6	90
Cod. Vinob. 10825, 181r	C93, 205r	Halberd 7	92
Cod. Vinob. 10825, 181v	C93, 205v	Halberd 8	94
Cod. Vinob. 10825, 182r	C93, 206r	Halberd 9	96
Cod. Vinob. 10825, 182v	C93, 206v	Halberd 10	98
Cod. Vinob. 10825, 183r	C93, 207r	Halberd 11	100
Cod. Vinob. 10825, 183v	C93, 207v	Halberd 12	102
Cod. Vinob. 10825, 184r	C93, 208r	Halberd 13	104
Cod. Vinob. 10825, 184v	C93, 208v	Halberd 14	106
Cod. Vinob. 10825, 185r	C93, 209r	Halberd 15	108
Cod. Vinob. 10825, 185v	C93, 209v	Halberd 16	110
Cod. Vinob. 10825, 186r	C93, 210r	Halberd 17	112
Cod. Vinob. 10825, 186v	C93, 210v	Halberd 18	114
Cod. Vinob. 10825, 187r	C93, 211r	Halberd 19	116
Cod. Vinob. 10825, 187v	C93, 211v	Halberd 20	118
Cod. Vinob. 10826, 120r	—	Halberd 21*	120

Vienna	**Dresden**	**Plate**	**Page**
Cod. Vinob. 10826, 149r	C94, 164r	Poleax 1	124
Cod. Vinob. 10826, 149v	C94, 164v	Poleax 2	126
Cod. Vinob. 10826, 150r	C94, 165r	Poleax 3	128
Cod. Vinob. 10826, 150v	C94, 165v	Poleax 4	130
Cod. Vinob. 10826, 151r	C94, 166r	Poleax 5	132
Cod. Vinob. 10826, 151v	C94, 166v	Poleax 6	134
Cod. Vinob. 10826, 152r	C94, 167r	Poleax 7	136
Cod. Vinob. 10826, 152v	C94, 167v	Poleax 8	138
Cod. Vinob. 10826, 153r	C94, 168r	Poleax 9	140
Cod. Vinob. 10826, 153v	C94, 168v	Poleax 10	142

Vienna	Dresden	Plate	Page
Cod. Vinob. 10826, 154r	C94, 169r	Poleax 11	144
Cod. Vinob. 10826, 154v	C94, 169v	Poleax 12	146
Cod. Vinob. 10826, 155r	C94, 170r	Poleax 13	148
Cod. Vinob. 10826, 155v	C94, 170v	Poleax 14	150
Cod. Vinob. 10826, 156r	C94, 171r	Poleax 15	152
Cod. Vinob. 10826, 156v	C94, 171v	Poleax 16	154

Vienna	Dresden	Plate	Page
Cod. Vinob. 10826, 15r	C93, 227r	Various 1	158
Cod. Vinob. 10826, 15v	C93, 227v	Various 2	160
Cod. Vinob. 10826, 13r	C93, 228r	Various 3	162
Cod. Vinob. 10826, 13v	C93, 228v	Various 4	164
Cod. Vinob. 10826, 19r	C93, 232r	Various 11	166
Cod. Vinob. 10826, 19v	C93, 232v	Various 12	168
Cod. Vinob. 10826, 120v	–	Various 22*	170
Cod. Vinob. 10826, 121r	–	Various 23*	172
Cod. Vinob. 10826, 121v	–	Various 24*	174

* Images marked with an asterisk are composites created by author David James Knight.

BIBLIOGRAPHY

We have included resources for further study and research as well as works cited. For updates, articles and training materials supplementary to this volume, please visit our official website at **http://www.PaulusHectorMair.com**.

Andrews, William. *Old Time Punishments*. New York: Dorset Press, 1991.

Betts, Gavin. *Teach Yourself Latin*. Chicago: Hodder & Stoughton Ltd., 2003.

Birnbaum, Norman. *Social Structure and the German Reformation*. New York: Arno Press, 1980.

Borchardt, Frank L. *German Antiquity in Renaissance Myth*. Baltimore: Johns Hopkins Press, 1971.

Cappelli, Adriano. *Dizionario di Abbreviature Latine ed Italiane*. 6th ed. Milano, Ulrico Hoepli, 2001.

Clements, John. "A Brief Look at Grappling and Wrestling in Renaissance Fencing." *Spada: An Anthology of Swordsmanship*. Ed. Stephen Hand. Union City: Chivalry Bookshelf, 2002.

Durrell, Martin, Katrin Kohl, and Gudrun Loftus. *Essential German Grammar*. London: McGraw Hill, 2002.

Grun, Bernard. *The Timetables of History*. 3rd ed. New York: Simon & Schuster, 1991.

Hale, J. R. *War and Society in Renaissance Europe, 1450–1620*. Avon: Leicester University Press, 1985.

Hall, Bert S. *Weapons and Warfare in Renaissance Europe*. Baltimore: Johns Hopkins University Press, 1997.

Harrington, Karl. *Mediaeval Latin*. Norwood: Allyn and Bacon, 1925.

Hennig, Beate. *Leines Mittelhochdeutsches Wörterbuch*. Tübingen: Max Niemeyer, 2001.

Hils, Hans-Peter. *Meister Johann Liechtenauers Kunst des Langen Schwertes*. Diss. Universität Freiburg. Frankfurt: Peter Lang, 1985.

Kidd, D. A. *Collins Gem Latin Dictionary*. Ed. Joyce Littlejohn. 2nd ed. Glasgow: HarperCollins, 2003.

Klatt, E., D. Roy, G. Klatt, and H. Messinger. *Langenscheidt's Standard German Dictionary*. New York: Langenscheidt, 1983.

Lau, Franz, and Ernst Bizer. *A History of the Reformation in Germany*. London: Adam & Charles Black, 1969.

Lewis, Charlton, and Charles Short. *A Latin Dictionary*. "The Persius Project." Tufts University. http://www.perseus.tufts.edu/cgi-bin/resolveform?lang=Latin.

Lindholm, David, and Peter Svärd. *Sigmund Ringeck's Knightly Art of the Longsword*. Boulder: Paladin Press, 2003.

Mair, Paulus H. Cod. Vinob. 10825 & 10826. *Opus Amplissimum de Arte Athletica*. Österreichische Nationalbibliothek, Vienna.

Mair, Paulus H. C93 & C94. *Fecht-, Ring-, und Turnierbuch*. Sächsische Landesbibliothek, Dresden.

Mair, Paulus H. Cod. Icon. 393. Bayerische Staatsbibliothek, Munich.

Moeller, Bernd. *Imperial Cities and the Reformation*. Trans. H. C. Erik Midelfort. Durham: Labyrinth Press, 1982.

Oakeshott, Ewart. *A Knight and His Armor*. 2nd ed. Chester Springs: Dufour Editions, 1999.

Oakeshott, Ewart. *A Knight and His Weapons*. 2nd ed. Chester Springs: Dufour Editions, 1997.

Oakeshott, Ewart. *A Knight in Battle*. 2nd ed. Chester Springs: Dufour Editions, 1998.

Oakeshott, Ewart. *The Archaeology of Weapons*. Woodbridge, U.K.: Boydell Press, 1964.

Prior, Richard, and Joseph Wohlberg. *501 Latin Verbs*. New York: Barron, 1995.

Simpson, D. P. *Cassell's Latin Dictionary*. 5th ed. New York: John Wiley, 1968

Terrell, Peter, Veronkica Schnorr, Wendy Smith, and Roland Breitsprecher. *Collins Dictionary Deutsch to English*. 5th ed. New York: HarperCollins, 2004.

Traupman, John C. *The New College Latin & English Dictionary*. New York: Bantam Books, 1995.

Wheelock, Frederic M. *Wheelock's Latin*. Rev. Richard A. Lafleur. 6th ed. New York: HarperCollins, 2000.

Whitaker, William. "Words Latin to English Dictionary." http://ablemedia.com/ctcweb/showcase/whitakerwords.html.

Zabinski, Grzegorz, and Bartlomiej Walczak. *Codex Wallerstein*. Boulder: Paladin Press, 2002.

ABOUT THE AUTHORS

David James Knight is a former Army officer, Iraq veteran, and lifelong martial artist. He holds a B.A. magna cum laude from Florida International University, where he focused on medieval literature and Latin, and a J.D. from Washington and Lee University.

Brian Hunt's interest in martial arts began at a very young age. In order to study various untranslated works, he taught himself to read Latin, German and Italian. Other translations by him include the MS I:33 and Hans Talhoffer's 1459 manuscript. His lifelong interest in martial arts and fencing led him into many areas of study, including various Japanese and Chinese styles (with an emphasis on Okinawan traditions) along with unarmed and armed Western martial arts both in and out of armored harness.

www.ingramcontent.com/pod-product-compliance
Lightning Source LLC
Chambersburg PA
CBHW040845100426
42812CB00014B/2613